Edited by
Wolfgang Jorisch

Vacuum Technology in the Chemical Industry

Related Titles

Tsotsas, E., Mujumdar, A. S. (eds.)

Modern Drying Technology

5 Volume Set

2014
ISBN 978-3-527-31554-3

Jousten, K. (ed.)

Handbook of Vacuum Technology

2008
Print ISBN: 978-3-527-40723-1

Bannwarth, H.

Liquid Ring Vacuum Pumps, Compressors and Systems

Conventional and Hermetic Design

2005
Print ISBN: 978-3-527-31249-8

Oetjen, G., Haseley, P.

Freeze-Drying

2 Edition

2004
Print ISBN: 978-3-527-30620-6

O'Hanlon, J.F.

A User's Guide to Vacuum Technology, 3rd Edition

3 Edition

2003
Print ISBN: 978-0-471-27052-2

Zhu, W. (ed.)

Vacuum Microelectronics

2001
Print ISBN: 978-0-471-32244-3

Lafferty, J.M. (ed.)

Foundations of Vacuum Science and Technology

1998
Print ISBN: 978-0-471-17593-3

Edited by Wolfgang Jorisch

Vacuum Technology in the Chemical Industry

Verlag GmbH & Co. KGaA

Editor

Dr. Wolfgang Jorisch
Kreisbahnstrasse 66
52511 Geilenkirchen
Germany

■ All books published by **Wiley-VCH** are carefully produced. Nevertheless, authors, editors, and publisher do not warrant the information contained in these books, including this book, to be free of errors. Readers are advised to keep in mind that statements, data, illustrations, procedural details or other items may inadvertently be inaccurate.

Library of Congress Card No.: applied for

British Library Cataloguing-in-Publication Data
A catalogue record for this book is available from the British Library.

Bibliographic information published by the Deutsche Nationalbibliothek
The Deutsche Nationalbibliothek lists this publication in the Deutsche Nationalbibliografie; detailed bibliographic data are available on the Internet at <http://dnb.d-nb.de>.

© 2015 Wiley-VCH Verlag & Co. KGaA, Boschstr. 12, 69469 Weinheim, Germany

All rights reserved (including those of translation into other languages). No part of this book may be reproduced in any form – by photoprinting, microfilm, or any other means – nor transmitted or translated into a machine language without written permission from the publishers. Registered names, trademarks, etc. used in this book, even when not specifically marked as such, are not to be considered unprotected by law.

Print ISBN: 978-3-527-31834-6
ePDF ISBN: 978-3-527-65392-8
ePub ISBN: 978-3-527-65391-1
Mobi ISBN: 978-3-527-65390-4
oBook ISBN: 978-3-527-65389-8

Cover Design Adam Design
Typesetting Laserwords Private Limited, Chennai, India
Printing and Binding STRAUSS GmbH, Mörlenbach, Germany

Printed on acid-free paper

Contents

List of Contributors *XV*

1   Fundamentals of Vacuum Technology *1*
    Wolfgang Jorisch
1.1   Introduction *1*
1.2   Fundamentals of Vacuum Technology *2*
1.2.1   Fundamentals of Gas Kinetics *3*
1.2.1.1   Mean Free Path *5*
1.2.2   Equation of State for Ideal Gases *6*
1.2.3   Flow of Gases through Pipes in a Vacuum *7*
1.2.3.1   Gas Throughput and Conductance *8*
1.2.3.2   Flow through Long, Round Pipes *10*
1.2.4   Vacuum Pumps Overview *12*
    References *14*

2   Condensation under Vacuum *15*
    Harald Grave
2.1   What Is Condensation? *15*
2.2   Condensation under Vacuum without Inert *16*
2.3   Condensation with Inert Gases *17*
2.4   Saturated Inert Gas–Vapour Mixtures *19*
2.5   Vapour–Liquid Equilibrium *20*
2.6   Types of Condensers *21*
2.7   Heat Transfer and Condensation Temperature in a Surface Condenser *24*
2.8   Vacuum Control in Condensers *30*
2.9   Installation of Condensers *30*
2.10   Special Condenser Types *32*
    Further Reading *34*

3   Liquid Ring Vacuum Pumps in Industrial Process Applications *35*
    Pierre Hähre
3.1   Design and Functional Principle of Liquid Ring Vacuum Pumps *35*

3.1.1   Functional Principle   35
3.1.2   Design Details   37
3.1.2.1   The Medium Conveyance   37
3.1.2.2   Design of the Pressure Ports   37
3.1.2.3   Situation of the Working Openings   39
3.2   Operating Behaviour and Design of Liquid Ring Vacuum Pumps   40
3.2.1   Hydraulics   40
3.2.1.1   Power Consumption   40
3.2.1.2   Suction Volume   41
3.2.2   Thermodynamics   41
3.2.2.1   Condensable Components in the Medium to Be Pumped   42
3.2.2.2   Temperature of the Medium to Be Pumped   42
3.2.2.3   Operating Liquid   43
3.2.2.4   Catalogue Values   44
3.2.3   Counterpressure, Air Pressure   45
3.2.4   Design Data   47
3.3   Vibration and Noise Emission with Liquid Ring Vacuum Pumps   48
3.3.1   Vibration Stimulation by Imbalance of Rotary Solids   48
3.3.2   Vibration Stimulation by Pulsation   49
3.3.3   Vibration Stimulation by Flow Separations   50
3.3.4   Measures for Vibration Damping   50
3.3.4.1   Factory Measures   50
3.3.4.2   Installation   50
3.3.4.3   Operating Mode   51
3.4   Selection of Suitable Liquid Ring Vacuum Pumps   51
3.4.1   Simple, Robust and Suitable for the Entire Pressure Range   52
3.4.2   The Vacuum Pump for the Delivery of Liquid   52
3.4.3   Quiet and Compact for a Vacuum Close to the Vapour Pressure   57
3.4.4   A Compact Unit   59
3.4.5   The Right Sealing Concept   59
3.4.5.1   Single-Acting Mechanical Seal   59
3.4.5.2   Double-Acting Mechanical Seal   61
3.4.5.3   Magnetic Coupling   62
3.4.6   Vacuum Control   63
3.4.7   Valve Control   63
3.4.7.1   Bypass Recovery   63
3.4.7.2   Reduction on the Suction Side   64
3.4.8   Power Adjustment   64
3.4.8.1   Cascade Connection of Vacuum Pumps   65
3.4.8.2   Speed Adjustment in Vacuum Pumps   65
3.4.8.3   Cascade Connection Combined with Speed Adjustment   66
3.5   Process Connection and Plant Construction   67
3.5.1   Set-Up and Operation of Liquid Ring Vacuum Pumps   67
3.5.2   Conveyance of the Operating Liquid   68
3.5.3   Precompression   69

3.5.3.1	Precompression by Gas Ejector	69
3.5.3.2	Precompression by Steam Jet	71
3.5.3.3	Precompression by Rotary Piston Vacuum Pumps	72
3.6	Main Damage Symptoms	73
3.6.1	Water Impact	73
3.6.2	Cavitation	73
3.6.2.1	Concept	73
3.6.3	Calcareous Deposits and How to Avoid Them	75
3.6.3.1	Water as Service Liquid in Liquid Ring Vacuum Pumps (LRVPs)	75
3.6.3.2	Prevention by Monitoring the Water Quality	76
3.7	Table of Symbols	78

4 Steam Jet Vacuum Pumps 81
Harald Grave

4.1	Design and Function of a Jet Pump	81
4.2	Operating Behaviour and Characteristic	84
4.3	Control of Jet Compressors	87
4.4	Multi-Stage Steam Jet Vacuum Pumps	90
4.5	Comparison of Steam, Air and Other Motive Media	93
	Further Reading	95

5 Mechanical Vacuum Pumps 97
Wolfgang Jorisch

5.1	Introduction	97
5.2	The Different Types of Mechanical Vacuum Pumps	99
5.2.1	Reciprocating Piston Vacuum Pump	100
5.2.2	Diaphragm Vacuum Pump	100
5.2.3	Rotary Vane Vacuum Pump	101
5.2.4	Rotary Plunger Vacuum Pump	102
5.2.5	Roots Vacuum Pump	103
5.2.6	Dry Compressing Vacuum Pump	104
5.3	When Using Various Vacuum Pump Designs in the Chemical or Pharmaceutical Process Industry, the Following Must Be Observed	104
5.3.1	Circulatory-Lubricated Rotary Vane and Rotary Plunger Vacuum Pumps	104
5.3.2	Fresh-Oil-Lubricated Rotary Vane Vacuum Pumps	108
5.3.3	Dry, Respectively Oil-Free Compressing Vacuum Pumps	110
5.3.4	Roots Vacuum Pumps	111
5.3.4.1	Operating Principle of Roots Vacuum Pumps	111
5.3.4.2	Roots Vacuum Pumps with Bypass Valve, Respectively with Frequency Controlled Motor	112
5.3.4.3	Compression of Roots Vacuum Pumps	113
5.3.4.4	Dimensioning Combinations of Roots Vacuum Pumps with Backing Pumps	115

5.3.4.5	Power Requirement of a Roots Vacuum Pump	*117*
5.3.4.6	Roots Vacuum Pumps with Pre-Admission Cooling Facility	*117*
5.3.5	Dry Compressing Vacuum Pumps for Chemistry Applications	*118*
5.3.5.1	Dry Compressing, Three-Stage Roots Vacuum Pump with Exhaust, Respectively Non-Return Valves between the Stages	*118*
5.3.5.2	Claw Vacuum Pumps (Northey Principle)	*119*
5.3.5.3	New Developments of Screw Vacuum Pumps for the Area of Process Chemistry	*124*
5.3.5.4	Outlook as to the Future of the Mechanical Vacuum Pumps in the Area of Chemical Process Engineering	*128*
	References	*128*

6 Basics of the Explosion Protection and Safety-Technical Requirements on Vacuum Pumps for Manufacturers and Operating Companies *129*
Hartmut Härtel

6.1	Introduction	*129*
6.2	Explosion Protection	*130*
6.2.1	General Basics	*130*
6.2.2	Explosive Atmosphere and Safety Characteristics	*131*
6.2.2.1	General	*131*
6.2.2.2	Explosion Range and Explosion Limits	*133*
6.2.2.3	Flash Point	*137*
6.2.2.4	Minimum Ignition Energy and Ignition Temperature	*138*
6.2.2.5	Flame-Proof Gap Width	*141*
6.2.3	Measures of the Explosion Protection	*144*
6.3	Directive 99/92/EC	*146*
6.3.1	Requirements on Operating Companies of Vacuum Pumps	*146*
6.3.2	Classification of Hazardous Areas into Zones	*149*
6.4	Directive 94/9/EC	*150*
6.4.1	Equipment Groups and Categories	*150*
6.4.1.1	Equipment Group II	*151*
6.4.2	Assignment between Equipment Categories and Zones	*152*
6.4.3	Requirements on Manufacturers of Vacuum Pumps	*153*
6.4.4	Conformity Assessment Procedure	*154*
6.4.5	Application of the Regulations of the Directive	*155*
6.5	Summary	*157*
	References	*158*
	Further Reading	*159*

7 Measurement Methods for Gross and Fine Vacuum *161*
Werner Große Bley

7.1	Pressure Units and Vacuum Ranges	*161*
7.2	Directly and Indirectly Measuring Vacuum Gauges and Their Measurement Ranges	*162*
7.3	Hydrostatic Manometers	*163*

7.4	Mechanical and Electromechanical Vacuum Gauges *164*
7.4.1	Sensors with Strain Gauges *165*
7.4.2	Thermal Conductivity Gauges *167*
7.4.3	Thermal Conductivity Gauges with Constant Filament Heating Power *169*
7.4.4	Thermal Conductivity Gauges with Constant Filament Temperature *170*
7.4.5	Environmental and Process Impacts on Thermal Conductivity Gauges *170*
	References *172*
	Further Reading *172*
8	**Leak Detection Methods** *173*
	Werner Große Bley
8.1	Definition of Leakage Rates *173*
8.2	Acceptable Leakage Rate of Chemical Plants *174*
8.3	Methods of Leak Detection *175*
8.4	Helium as a Tracer Gas *176*
8.5	Leak Detection with Helium Leak Detector *176*
8.6	Leak Detection of Systems in the Medium-Vacuum Range *177*
8.6.1	Connection of Leak Detector to the Vacuum System of a Plant *177*
8.6.2	Detection Limit for Leakage Rates at Different Connection Positions of a Multistage Pumping System *179*
8.7	Leak Detection on Systems in the Rough Vacuum Range *180*
8.7.1	Connection of Leak Detector Directly to the Process Vacuum *180*
8.7.2	Connection of Leak Detector at the Exhaust of the Vacuum System *180*
8.8	Leak Detection and Signal Response Time *181*
8.9	Properties and Specifications of Helium Leak Detectors *182*
8.10	Helium Leak Detection in Industrial Rough Vacuum Applications without Need of a Mass Spectrometer *183*
8.10.1	Principle of the Wise Technology® Sensor *185*
8.10.2	Application *186*
	References *187*
	Further Reading *187*
	European Standards *187*
9	**Vacuum Crystallisation** *189*
	Guenter Hofmann
9.1	Introduction *189*
9.2	Crystallisation Theory for Practice *189*
9.3	Types of Crystallisers *195*
9.4	Periphery *203*
9.5	Process Particularities *205*
9.5.1	Surface-Cooling Crystallisation *206*

9.5.2	Vacuum-Cooling Crystallisation 207
9.5.3	Evaporation Crystallisation 207
9.6	Example – Crystallisation of Sodium Chloride 207
	References 209

10 Why Evaporation under Vacuum? 211
Gregor Klinke

Summary 211

- 10.1 Introduction 211
- 10.2 Thermodynamics of Evaporation 211
- 10.3 Pressure/Vacuum Evaporation Comparison 213
- 10.3.1 Vapour Utilisation 214
- 10.3.2 Design of the Apparatuses 214
- 10.3.3 Machine Equipment 214
- 10.3.4 Corrosion 215
- 10.3.5 Insulation 215
- 10.3.6 Safety Aspects 215
- 10.3.7 Product Properties 215
- 10.3.8 Boiling Range 216
- 10.4 Possibility of Vapour Utilization 217
- 10.4.1 External Utilization 217
- 10.4.2 Multi-Stage Evaporation 217
- 10.4.3 Mechanical Vapour Recompression 217

Further Reading 220

11 Evaporators for Coarse Vacuum 221
Gregor Klinke

Summary 221

- 11.1 Introduction 221
- 11.2 Criteria for the Selection of an Evaporator 221
- 11.2.1 Suitability for the Product 221
- 11.2.2 Cleaning 222
- 11.2.3 Quality of Heat Transfer 222
- 11.2.4 Required Space 222
- 11.2.5 Cost Efficiency 223
- 11.3 Evaporator Types 223
- 11.3.1 Agitator Evaporator 223
- 11.3.2 Natural Circulation Evaporator 223
- 11.3.3 Climbing-Film Evaporator 225
- 11.3.4 Falling-Film Evaporator 226
- 11.3.5 Forced-Circulation Evaporator 228
- 11.3.6 Fluidised-Bed Evaporator 230
- 11.3.7 Plate Evaporator 231

Further Reading 233

12	**Basics of Drying Technology** *235*	
	Jürgen Oess	
12.1	Basics of Solids–Liquid Separation Technology *235*	
12.2	Basics of Drying Technology *235*	
12.2.1	Convection Drying *236*	
12.2.2	Radiation Drying *237*	
12.2.3	Contact Drying *237*	
12.2.3.1	Heat Transfer during Contact Drying *237*	
12.2.3.2	Product Temperature and Vapour Removal *239*	
12.2.3.3	Drying under Vacuum *241*	
12.2.4	Advantages of the Vacuum Drying *242*	
12.2.4.1	Increase of the Drying Capacity *242*	
12.2.4.2	Gentle Thermal Product Treatment *242*	
12.2.4.3	Separation of High Boiling Solvents *243*	
12.2.4.4	High Thermal Efficiency *243*	
12.2.4.5	Processing of Toxic or Explosive Materials *243*	
12.3	Discontinuous Vacuum Drying *244*	
12.3.1	Setup of a Batch Vacuum Drying System *244*	
12.3.2	Operation of Discontinuous Vacuum Dryers *244*	
12.4	Continuous Vacuum Drying *246*	
12.4.1	Setup of a Batch Vacuum Drying System *246*	
12.4.2	Operation of Continuous Vacuum Dryers *246*	
12.4.3	Inlet- and Outlet Systems *247*	
12.5	Dryer Designs *248*	
	Reference *249*	
13	**Vacuum Technology Bed** *251*	
	Michael Jacob	
13.1	Introduction to Fluidized Bed Technology *251*	
13.1.1	Open or Once-through Fluidized Bed Plants *251*	
13.1.2	Normal Pressure Fluidized Bed Units with Closed-Loop Systems *251*	
13.2	Vacuum Fluidized Bed Technology *253*	
13.2.1	Layout *253*	
13.2.2	Sequence of Operation *255*	
13.2.3	Fluidization at Vacuum Conditions *255*	
13.2.4	Heat Energy Transfer under Vacuum Conditions *256*	
13.2.5	Applications *257*	
	References *258*	
14	**Pharmaceutical Freeze-Drying Systems** *259*	
	Manfred Heldner	
14.1	General Information *259*	
14.2	*Phases* of a Freeze-Drying *Process* *260*	
14.2.1	Freezing *260*	

14.2.2	Primary Drying – Sublimation	261
14.2.3	Secondary Drying	264
14.2.4	Final Treatment	264
14.2.5	Process Control	265
14.3	Production Freeze-Drying Systems	266
14.3.1	Drying Chamber and Shelf Assembly	267
14.3.2	Ice Condenser	270
14.3.3	Refrigerating System	271
14.3.4	Vacuum System	273
14.3.5	Cleaning of the Freeze-Drying System	274
14.3.6	Sterilisation	276
14.3.7	VHP Sterilisation	277
14.4	Final Comments	278
	Further Reading	279
15	**Short Path and Molecular Distillation**	**281**
	Daniel Bethge	
15.1	Introduction	281
15.2	Some History	281
15.2.1	Vacuum Distillation	282
15.2.2	Short Path Evaporator	285
15.2.3	The Vacuum System	286
15.2.4	Distillation Plant	288
15.2.5	Application Examples	289
15.2.6	New Developments	292
15.3	Outlook	293
	References	293
16	**Rectification under Vacuum**	**295**
	Thorsten Hugen	
16.1	Fundamentals of Distillation and Rectification	295
16.2	Rectification under Vacuum Conditions	298
16.3	Vacuum Rectification Design	302
16.3.1	Liquid and Gas Load	303
16.3.2	Pressure Drop	303
16.3.3	Separation Efficiency	303
16.4	Structured Packings for Vacuum Rectification	305
	Nomenclature, Applied Units	309
	Greek Symbols	310
	Subscripts and Superscripts	310
	References	310
17	**Vacuum Conveying of Powders and Bulk Materials**	**311**
	Thomas Ramme	
17.1	Introduction	311

17.2	Basic Theory	*312*
17.2.1	General	*312*
17.2.2	Typical Conditions in a Vacuum Conveying Line	*315*
17.2.2.1	Dilute Phase Conveying	*316*
17.2.2.2	Dense Phase Conveying	*316*
17.2.2.3	Plug Flow Conveying	*317*
17.3	Principle Function and Design of a Vacuum Conveying System	*318*
17.3.1	Multiple-Stage, Compressed-Air Driven Vacuum Generators	*319*
17.3.2	Conveying and Receiver Vessels	*322*
17.3.3	Filter Systems	*324*
17.4	Continuous Vacuum Conveying	*325*
17.5	Reactor- and Stirring Vessel Loading in the Chemical Industry	*325*
17.6	Conveying, Weighing, Dosing and Big-Bag Filling and Discharging	*330*
17.7	Application Parameters	*330*
	References	*330*

18 Vacuum Filtration – System and Equipment Technology, Range and Examples of Applications, Designs *331*
Franz Tomasko

18.1	Vacuum Filtration, a Mechanical Separation Process	*331*
18.1.1	On the Theory of Filtration and Significance of the Laboratory Experiment	*332*
18.1.2	Guide to Filter-Type Selection	*333*
18.2	Design of an Industrial Vacuum Filter Station	*335*
18.3	Methods of Continuous Vacuum Filtration, Types of Design and Examples of Application	*337*
18.3.1	Vacuum Filtration on a Curved Convex Surface, the Drum Filter	*337*
18.3.1.1	Design of a Vacuum Drum Filter	*337*
18.3.1.2	Working Method of a Continuous Operating Vacuum Drum Filter	*345*
18.3.1.3	Different Constructions	*346*
18.3.1.4	Special Vacuum Drum Filters	*346*
18.3.1.5	Calculation Example	*348*
18.3.2	Vacuum Filtration on a Curved Concave Surface, the Internal Filter	*351*
18.3.3	Vacuum Filtration on a Flat Horizontal Surface	*352*
18.3.3.1	The Belt Filter	*352*
18.3.3.2	The Pan Filter	*353*
18.3.4	Vacuum Filtration on a Vertical Flat Surface, the Disc Filter	*358*
	References	*361*

Index *363*

List of Contributors

Daniel Bethge
GIG Karasek GmbH
Research and Development
Neusiedlerstraße 15-19
2640 Gloggnitz-Stuppach
Austria

Harald Grave
Forschung und Entwicklung
GAE Wiegand GmbH
Andreas Hofer Street 3
76185 Karlsruhe
Germany

Werner Große Bley
INFICON GmbH
Bonner Strasse 498
50968 Köln
Germany

Pierre Hähre
Speck Pumps GmbH & Co KG
Vacuum Technology Department
Regensburger Ring 6-8
91154 Roth
Germany

Hartmut Härtel
IBExU Institut für
Sicherheitstechnik GmbH
An-Institut der Bergakademie
Freiberg
Fuchsmühlenweg 7
09599 Freiberg
Germany

Manfred Heldner
Hüttengarten 23
53332 Bornheim-Widdig
Germany

Guenter Hofmann
GEA Messo PT Duisburg
Friedrich-Ebert-Street 134
47229 Duisburg
Germany

Thorsten Hugen
Julius Montz GmbH
Hofstraße 82
40723 Hilden
Germany

Michael Jacob
Glatt Ingenieurstechnik GmbH
Nordstr. 12
99427 Weimar
Germany

Wolfgang Jorisch
IVPT Industrielle
Vakuumprozeßtechnik
Kreisbahnstrasse 66
52511 Geilenkirchen
Germany

Gregor Klinke
GEA WIEGAND
Department Research and
Development
Andreas-Hofer-Straße 3
76185 Karlsruhe
Germany

Jürgen Oess
SiCor Engineering Partners Gbr
Burgunderweg 5
74357 Bönnigheim
Germany

Thomas Ramme
Volkmann GmbH
Schloitweg 17
59494 Soest
Germany

Franz Tomasko
FLSmidth Wiesbaden GmbH
Am Klingenweg 4a
65396 Walluf
Germany

1
Fundamentals of Vacuum Technology
Wolfgang Jorisch

1.1
Introduction

Vacuum technology is being used widely in many chemistry applications. Here it is not used in the same way as in physics applications. In physics applications, it is the objective to perform experiments in volumes (vacuum chambers) which are as pure as possible, that is, which contain as few particles as possible as these particles generally impair the physical process.

Vacuum technology is used in the area of chemistry applications for the purpose of performing basic thermal and mechanical operations to reprocess reaction products under conditions which preserve the product. Typical applications for thermal separating processes in a vacuum are distillation, drying or sublimation at reduced pressures as well as applications which accelerate the reaction itself when reaction products from the reaction mixture need to be removed for the purpose of shifting the equilibrium in the desired direction, for example. An example of this is the process of esterification.

A mechanical process performed in a vacuum is that of vacuum filtration where the pressure difference created between vacuum and atmospheric pressure is utilised as the driving force for the filtration process.

The planning process engineer or the consulting engineer of a chemical plant not only faces questions how to properly dimension a vacuum system so as to comply with the demanded process specifications, but he needs to solve in a satisfactory way, problems relating to operating costs which shall be as low as possible and questions as to the minimisation of emissions in the discharged air and waste water. The wide variety of vacuum pumps used in the area of chemistry technology reflects this. The responsible planning engineer or plant chemist will have to select, in consideration of the process engineering questions which differ from process to process, vacuum generators which promise to offer the best possible solution for the specific case.

For this reason, this book covers besides vacuum process engineering fundamentals, above all also the different types of vacuum pumps.

Vacuum Technology in the Chemical Industry, First Edition. Edited by Wolfgang Jorisch.
© 2015 Wiley-VCH Verlag GmbH & Co. KGaA. Published 2015 by Wiley-VCH Verlag GmbH & Co. KGaA.

1.2
Fundamentals of Vacuum Technology

Also for the vacuum technology used in chemistry applications, the underlying fundamental laws of physics apply.

The standard DIN 28400 Part 1 defines the vacuum state as

Vacuum is the state of a gas, the particle number density of which is below that of the atmosphere at the Earth's surface. Since the particle number density is within certain limits time and location dependent, it is not possible to state a general upper limit for the vacuum. Here the gas particles exert a pressure on all bodies which surround them, this pressure being the result of their temperature dependent motion. The pressure is defined as a force per unit of area, with the unit of measurement being the Pascal.

$$1\,\text{Pa} = 1\,\text{Nm}^{-2} \tag{1.1}$$

In the area of vacuum process engineering, frequently not Pascal is used as the unit of measurement for pressures but instead also the allowed unit 'bar' or 'mbar'.

$$1\,\text{mbar} = 10^2\,\text{Pa} = 1\,\text{hPa} \tag{1.2}$$

Owing to the differing behaviour of the particles within the considered volume (particle number density) which is dependent on the number of particles which are present, different pressure (vacuum) ranges have been defined with respect to their flow characteristics, for example:

	Pressure range (mbar)
Rough vacuum	$< 1.013\,10^3 - 1$
Medium vacuum	$< 1 - 10^{-3}$
High vacuum	$< 10^{-3} - 10^{-7}$
Ultrahigh vacuum	$< 10^{-7}$
Remark:	
New definition of beginning rough vacuum:	
Lowest pressure on Earth surface	300 mbar (Mount Everest)

The pressure When gas particles impinge on a wall (surface) they are subjected to a change in impulse, whereby they transfer an impulse to this wall. This impulse is the cause for the pressure exerted on the wall:

$$p = \frac{d(mv)}{A\,dt} = \frac{F}{A} \tag{1.3}$$

since the force is equivalent to the change in impulse of over time:

$$\frac{d(mv)}{dt} \tag{1.4}$$

p = pressure; F = force; A = area; m = mass; v = velocity; and t = time.

1.2 Fundamentals of Vacuum Technology

Table 1.1 Composition of atmospheric air.

Constituent	Volume share (%)	Partial pressure (mbar)
Nitrogen	78.09	780.9
Oxygen	20.95	209.5
Argon	0.93	9.3
Carbon dioxide	0.03	3.10^{-1}
Water vapour	≤2.3	≤ 23.3

Remainder: noble gases, hydrogen, methane, ozone, and so on.

When now considering a surface onto which particles impinge from a hemisphere and when integrating the transferred impulse over time, then one obtains for ideal gases

$$p = \frac{d(mv)}{Adt} = \bar{n}kT \tag{1.5}$$

$\bar{n}$ = particle number density; k = Boltzmann constant and T = absolute temperature.

That is, the exerted pressure is only dependent on the number of gas molecules n in the volume but is not dependent on the type of gas [1].

This statement ultimately confirms also Dalton's Law, which states that the total pressure of a gas atmosphere is equal to the sum of all partial pressures of this gas mixture:

$$p_{tot} = \bar{n}_{tot}kT = \bar{n}_1 kT + \bar{n}_2 kT + \cdots \bar{n}_i kT \tag{1.6}$$

or

$$p_{tot} = p_1 + p_2 + \cdots + p_i \tag{1.7}$$

p_{tot} = total pressure; $\bar{n}_{tot}$ = particle number density of all gas particles (types); and $\bar{n}_i$ = particle number density of the particle type i.

Given as an example is in the following the composition of air at atmospheric pressure [2] (Table 1.1).

1.2.1 Fundamentals of Gas Kinetics

The individual molecules or gas particles contained in a volume are in constant motion and collide with each other. In doing so, the particles change their velocity each time they collide. From a statistical point of view, all velocities are possible but with differing probability.

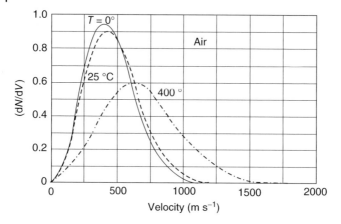

Figure 1.1 Velocity distribution of air molecules (nitrogen and oxygen) at 0, 25 and at 400 °C.

Maxwell and Boltzmann found the following relationship for the velocity distribution of the gas particles [3]:

$$\frac{dn}{dv} = \frac{2N}{\pi^{\frac{1}{2}}} \left(\frac{m}{2kT} \right)^{\frac{3}{2}} v^2 e^{\frac{-mv^2}{2kT}} \tag{1.8}$$

m	=	mass of each particle
T	=	temperature in Kelvin
N	=	number of particles
k	=	Boltzmann constant
v	=	velocity of the particles.

Figure 1.1 depicts the velocity distribution between velocity v and $v + dv$ based on the example of air at 0, 25 and 400 °C [3].

From this, it is apparent that there does not exist a molecule with a velocity of 'zero' or with an infinitely high velocity. The location of the most probable velocity (maximum, v_p) is a function of the mean gas temperature. Moreover, the molecule velocity depends on the molar mass. The most likely velocity can be stated through

$$v_p = \left(\frac{2kT}{m} \right)^{\frac{1}{2}} \tag{1.9}$$

and the arithmetic mean velocity can be stated through

$$v = \left(\frac{8kT}{\pi m} \right)^{\frac{1}{2}} \tag{1.10}$$

Figure 1.2 depicts the velocity of the gas molecules as a function of the type of gas.

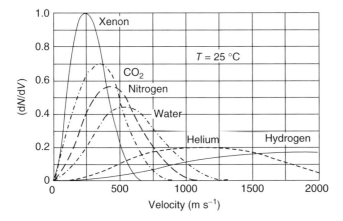

Figure 1.2 Gas molecule velocity as a function of the type of gas.

1.2.1.1 Mean Free Path

The fact that the molecules move at different velocities allows for the conclusion that they will move within a specific unit of time over a different distance (free path) before colliding with another particle. The mean free path λ, resulting from the kinetic gas theory is

$$\lambda = \frac{1}{2^{\frac{1}{2}} \pi d_0^2 \bar{n}} \tag{1.11}$$

where d_0 = collision radius in [mm] of an ideal point-like particle and $\bar{n}$ is the particle number density (number of molecules per volume).

Since the particle number density, as already derived, depends on the pressure, also the mean free path of the gas molecules is pressure dependent (at constant temperature), the product of the prevailing pressure and the mean free path is, at a given temperature, a constant (gas-type dependent).

For nitrogen at 20 °C, this product amounts to $\lambda \cdot p = 6.5 \cdot 10^{-5}$ mbar, that is at a prevailing pressure of $p = 1 \times 10^{-2}$ mbar there results at a temperature of 20 °C a mean free path of 6.5 mm [4].

Moreover, kinetic gas theory states the distribution of the mean free paths as follows:

$$N = N' e^{-\frac{x}{\lambda}} \tag{1.12}$$

N'	=	number of molecules in the volume
N	=	number of the molecules which have a free path of x.

From this equation, it can be derived that 63% of all collisions occur after a path of $0 \leq x \leq \lambda$ whereas 37% of the collisions occur after a path of $\lambda < x \leq 5\lambda$.

Only 0.6% have a free path exceeding 5λ.

1.2.2
Equation of State for Ideal Gases

In the year 1662, Robert Boyle demonstrated that the volume of a gas, provided it is maintained at a constant temperature, is inversely proportional to the pressure:

$$p_1 V_1 = p_2 V_2 \ (N, T \text{ constant(isotherm)}) \ Boyle's\,Law \tag{1.13}$$

In 1787, the French chemist Charles found the law according to which the volume of a gas is proportional to its temperature when maintaining it at a constant pressure:

$$\frac{V_1}{T_1} = \frac{V_0}{T_0} (N, p \text{ constant(isobar)}) \tag{1.14}$$

$$V_1 = V_0 \frac{273.15 + \vartheta}{273.15} \ Charles'\,Law$$
$$\vartheta = \text{temperature in }°C \tag{1.15}$$

Also, the French physicist Amontons discovered that the pressure of a gas is proportional to its temperature when maintaining it at a constant volume:

$$\frac{p_1}{T_1} = \frac{p_0}{T_0} (N, V \text{ constant(isochore)}) \tag{1.16}$$

$$p_1 = p_0 \frac{273,15 + \vartheta}{273,15} \ Amontons'\,Law \tag{1.17}$$

It is apparent for formal reasons that at $\vartheta = -273.15\,°C$ also the volume and the pressure of the ideal gas disappear. From this it can be concluded that the temperature $\vartheta = -273.15\,°C$ represents a physical limit for the temperature (absolute zero).

The laws of Charles and Amontons can be combined as

$$V = \frac{V_0 T}{T_0} \ (\text{Charles}) \tag{1.18}$$

and

$$p = \frac{p_0 T}{T_0} \ (\text{Amontons}) \tag{1.19}$$

One then obtains an equation of state which contains all state variables (p, V, T):

$$\frac{pV}{T} = \frac{p_0 V_0}{T} = \text{constant} \tag{1.20}$$

The Italian physicist Avogadro discovered in 1811, that the pressure of a gas is proportional to the number of molecules present:

$$\frac{p_1}{N_1} = \frac{p_2}{N_2} Avogadro's\,Law \ (T, V \text{ constant}) \tag{1.21}$$

At a 'standard temperature' of (0 °C) and a 'standard pressure' of (101 300 Pa) 1 mol of a gas always contains $N = 6.02252 \cdot 10^{23}$ particles taking up the volume of (V_{molar}) 22.4136 l (molar volume).

When inserting for p the 'standard pressure p_n', for T the 'standard temperature T_n,' and for V the 'molar volume V_{molar}', then one obtains the 'universal or molar gas constant R' for ideal gases:

$$R = \frac{p_n V_{molar}}{T_n} = \frac{101300 \text{Pa } 22.414 \text{ m}^3}{273.15 \text{ K kmol}} = 8.314 \frac{\text{kJ}}{\text{kmol K}} \quad (1.22)$$

or

$$R = 83.14 \frac{\text{mbar L}}{\text{mol K}} \quad (1.23)$$

Since the molar volume can also be expressed as the volume of the gas divided by the number of moles (n):

$$V_{molar} = \frac{V}{n} \quad (1.24)$$

one obtains from Eqs. (1.20) and (1.22) the universal equation of state for ideal gases:

$$pV = nRT \quad (1.25)$$

This equation of state which in effect only applies to 'ideal gases' may in a vacuum be applied with good accuracy also to real gases and vapours, because in vacuum the interactions between the particles reduce owing to the increasing free path. For this reason, this equation of state can be used to calculate pumped gas, respectively vapour quantities as they occur in connection with chemical engineering processes.

1.2.3
Flow of Gases through Pipes in a Vacuum

The flow conditions within a gas can be described through the Knudsen number. Flows through pipes are characterised through the Reynolds number. At relatively high pressures in a rough vacuum, a viscous flow is present which may either be laminar or turbulent. For a viscous flow, a mutual influence between the flowing particles is typical. In the turbulent range, the molecules behave chaotically. The Knudsen number is defined as follows:

$$Kn = \frac{\lambda}{d} \quad (1.26)$$

A viscous gas flow is characterised through a Knudsen number of $Kn < 0.01$.

The pipe diameter d is here much bigger than the mean free path λ of the molecules, and the gas flow is characterised by constant collisions amongst the particles.

The boundary between a continuous flow and a turbulent flow can be characterised through the Reynolds number (Re):

$$Re = \frac{u\rho d}{\eta} \quad (1.27)$$

ρ	=	gas density in kg/m³
η	=	viscosity the gas
u	=	flow velocity
d	=	pipe diameter.

The following applies to the flow through pipes:

When the non-dimensional Reynolds number attains values of over 2200, then one speaks of a turbulent flow, in the case of values below 1200 one then speaks of a laminar flow. In the range in between, either turbulent or laminar flow conditions can be present.

When the mean free path λ of the molecules is equal to the pipe diameter d or exceeds it, then there results a Knudsen number of >1 and when the Reynolds number is <1200 then the behavior of the gas molecules becomes more and more individual, that is, the flow changes to a 'molecular flow'.

1.2.3.1 Gas Throughput and Conductance

The term gas throughput is defined as a quantity of gas flowing through a given area per unit of time.

The gas quantity Q corresponds here to the volume ($\dot{V}$) of gas occurring per unit of time multiplied with the prevailing pressure p and Eq. (1.29):

$$Q = p\dot{V} \tag{1.28}$$

The SI unit for the gas throughput Q is Pa m³ s⁻¹.

The gas throughput corresponds to an energy passing through an area per unit of time [3]

$$1 \, \text{Pa} \, \text{m}^3 \, \text{s}^{-1} = 1 \, \text{W} \tag{1.29}$$

From Eq. (1.28) one then obtains the gas throughput as a mass flow:

$$\dot{m} \, \text{kg} \, \text{s}^{-1} = \frac{QM_{\text{molar}}}{RT} \tag{1.30}$$

M_{molar} = molar mass of the gas.

The transported gas quantity Q, which flows through a pipe is in the case of a continuous flow dependent on the prevailing pressure difference between inlet and outlet.

When taking the ratio between the gas quantity being throughput and the prevailing pressure difference across the pipe, then one obtains a quantity which can be designated as 'conductance' L of this pipe:

$$L = \frac{Q}{\Delta P} \tag{1.31}$$

Equation (1.31) is also termed 'Ohm's Law' of vacuum technology. When now considering the flow of a gas through an orifice (very short pipe) one obtains a rather complicated dependency of the gas being throughput as a function of the pressure difference ahead of, and after the orifice. Let the pressure ahead of the orifice be the atmospheric standard pressure, whereas the pressure is being reduced behind the orifice. Through the reduction in 'flow out pressure' the gas quantity being put through will increase until it attains a maximum. One now speaks of a critical pressure ratio, the gas then flows through the orifice at the speed of sound.

The gas quantity now passing through the orifice can be calculated from [3]

$$Q = Ap_1 C' \left(\frac{2\kappa kT}{(\kappa - 1)m} \right)^{\frac{1}{2}} \left(\frac{p_2}{p_1} \right)^{\frac{1}{\kappa}} \left[1 - \left(\frac{p_2}{p_1} \right)^{\frac{(\kappa-1)}{\kappa}} \right]^{\frac{1}{2}} \tag{1.32}$$

where $\kappa = C_p/C_V$, A = area of the orifice, p_1 = pressure ahead of the orifice, p_2 = pressure downstream of the orifice, C_p, C_V = heat capacity at constant pressure, respectively constant volume.

C' is a reduction factor since the gas which passes at a high velocity through the orifice, is subject to a volume contraction downstream of the orifice ('vena contracta').

For thin orifices, C' has a value of approximately 0.85.

When after attaining the speed of sound, the flow out pressure is reduced further, then the gas throughput will not increase further. Under these conditions the following applies

$$\frac{p_2}{p_1} = \left(\frac{2}{(\kappa + 1)} \right)^{\frac{\kappa}{\kappa-1}}$$

and one speaks of the critical expansion ratio.

The gas flow Q which is then being put through becomes

$$Q = Ap_1 C' \left(\frac{kT}{m} \frac{2\kappa}{\kappa + 1} \right)^{\frac{1}{2}} \left(\frac{2}{\kappa + 1} \right)^{\frac{1}{(\kappa-1)}} \tag{1.33}$$

And is thus independent of the flow out pressure p_2.

Equation (1.33) represents the critical gas flow. For air, $\kappa = 1.4$.

And blocking takes place at $p_2/p_1 = 0.525$. Figure 1.3 clarifies the flow under blocking conditions.

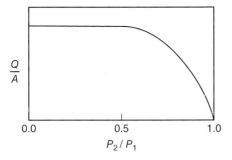

Figure 1.3 Gas flow under blocking conditions after attaining the critical pressure ratio.

These flow resistance considerations are important to vacuum systems when through these, gas flows are restricted in connection with pump-down processes or venting processes or when measuring leaks, for example.

1.2.3.2 Flow through Long, Round Pipes

The most well-known mathematical approach describing a viscous flow through long, straight pipes is through the so-called *Hagen–Poiseuille equation*:

$$Q = \frac{\pi d^4}{128 \eta l} \frac{p_1 + p_2}{2} (p_1 - p_2) \tag{1.34}$$

The conductance L of a cylindrical pipe for air at 0 °C is stated in litres per second whereby diameter d and pipe length l are entered in centimetres and the pressure p is entered in millibars as follows [4]:

$$L = 135 \frac{d^4}{l} \frac{p_1 + p_2}{2}$$

This special solution is valid, provided the following additional conditions are fulfilled:

- The flow profile in the pipe is fully present (location independent, applies to long pipes).
- The flow conditions are laminar, that is, $Re < 1200$ and $Kn < 0.01$
- The ratio between the actual gas velocity and speed of sound is <0.3.

A comprehensive description of the flow conditions in gas flows also through conducting sections exhibiting different geometries under different pressure and flow conditions can be found in [2, 3].

Analogously to 'Ohm's Law' the conductance changes when connecting pipes, valves or equipment like condensers as follows:

- Conductances add for a parallel connection:

$$L_{\text{tot}} = L_1 + L_2 + \cdots L_i \tag{1.35}$$

- Conductances add reciprocally for a series connection:

$$\frac{1}{L_{\text{tot}}} = \frac{1}{L_1} + \frac{1}{L_2} + \cdots \frac{1}{L_i} \tag{1.36}$$

Figure 1.4 states the conductances for pipes of commonly used nominal widths for a laminar flow [5].

If in the pipe, bends and components like valves are present then a correction is introduced by assuming a longer length of the pipe ('equivalent pipe lengths').

In Figure 1.5 the equivalent pipe lengths of valves and pipe bends are stated. As to the conductance values for equipment it will be necessary to ask the manufacturers for such information. As a rule, all vacuum lines should be as short as possible and have a diameter which is as large as possible (same dimension as the intake port on the vacuum pump used).

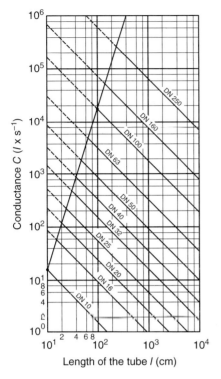

Figure 1.4 Conductance values of pipes of common nominal widths with a circular cross-section (apply to a laminar flow ($p = 1$ mbar), flowing medium is air).

DN	10	16	25	40	65	100	150	250
90° elbows; $D = 3\,d$	0.2	0.3	0.4	0.6	1.0	1.5	2.5	3.1
Straight-line valve	0.8	0.9	1.0	1.1	1.6	2.4	3.8	5.2
Right-angle valve	0.7	0.8	0.9	1.0	1.4	2.0	2.6	3.3

Figure 1.5 Equivalent pipe lengths in metres for components.

In the case of a molecular flow, the conductance becomes independent of the mean pressure difference across the length of the pipe [4]:

$$L = 12.1 \frac{d^3}{l} \text{ in litres per second} \qquad (1.37)$$

1.2.4
Vacuum Pumps Overview

The standard DIN 28400 Part 2 provides an overview of commonly used vacuum pump types.

However, in the area of chemical process engineering only pumps of the following general types are used:

- *Positive displacement pumps:* diaphragm vacuum pumps, liquid ring vacuum pumps, rotary vane and rotary piston vacuum pumps as well as Roots vacuum pumps (dry compressing claw and screw pumps are not yet mentioned here but also belong to this group of pumps).
- *Kinetic vacuum pumps:* vapour jet pumps, gas ejector pumps and liquid jet pumps.
- *Adsorption pumps:* the condenser.

The specific design types of different vacuum pumps, their special characteristics and application in the area of chemical process engineering are covered in the following chapters of this book.

The areas of application for these vacuum pumps are all within the rough and medium vacuum range, the principal vacuum range for chemistry processes. Only the short path and molecular distillation processes rely also high-vacuum pumps like the diffusion pump or even the turbomolecular pump (kinetic gas pumps). High-vacuum pumps are not covered in this book. For these refer to [2, 3].

It is the task of all vacuum pumps outlined above to generate a vacuum in a process engineering plant, for example, and to provide enough pumping speed. A pumping speed is in the end only a volume which passes per unit of time through an area.

Or the pumping speed S of a vacuum pump corresponds to the gas throughput Q, which passes through the area of the intake port of a vacuum pump, for example within a certain time ($\dot{Q}$), divided by the prevailing pressure.

$$S = \frac{\dot{Q}}{P} \tag{1.38}$$

The pumping speed S of a vacuum pump operating in the rough and medium vacuum range is commonly stated in $m^3 h^{-1}$.

Figure 1.6 depicts the so-called pumping speed diagram of a mechanical rotary vane vacuum pump. The pumping speed is graphed as a function of the intake pressure.

The evacuation process for a vacuum chamber, like that of a distillation column, is effected with a rough vacuum pump, regardless of type, as follows:

The gas quantity $-(V_K \Delta p)/\Delta t$ pumped out per unit of time Δt from a constant distillation column volume V_K flows at the prevailing pressure in the distillation column into the intake port of the vacuum pump. The corresponding gas quantity Q is equal to the effective pumping speed S_{eff} of the vacuum pump, respectively the pump system connected to the vacuum chamber (due to low conductance values

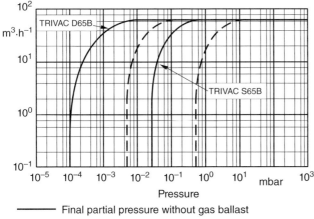

— Final partial pressure without gas ballast
- - - Final total pressure with gas ballast

Figure 1.6 Pumping speed diagram based on the example of a rotary vane vacuum pump.

of the vacuum line, the pumping speed of the vacuum pump may suffer throttling), multiplied by the prevailing pressure:

$$\frac{-V_K \Delta p}{\Delta t} = p S_{\text{eff}}$$

The change in pressure within the distillation column is proportional to the pump-down time:

$$\frac{\Delta p}{p} = -\frac{S_{\text{eff}}}{V_K} \Delta t \tag{1.39}$$

By integrating the relative change in column pressure over time, one obtains the relationship between pumpdown time t, the attained pressure p in the column and the initial pressure p_0 [6]:

$$t = \frac{V_K}{S_{\text{eff}}} \ln \frac{p_0}{p} = \frac{V_K}{S_{\text{eff}}} 2.30 \log \frac{p_0}{p}, \tag{1.40}$$

respectively

$$p = p_0 e^{-\frac{S_{\text{eff}}}{V_K} t} \tag{1.41}$$

This relationship applies to a pressure range, in which the pumping speed of the vacuum pump, respectively the pump system is almost constant (see Figure 1.6).

For vacuum applications in the area of chemical process engineering, the question as to a specific pumpdown time rarely arises. Instead process and leakage gas flows need to be pumped away by a vacuum pump to such an extent that a demanded operating pressure is reliably attained and maintained.

The questions relating to the evacuation time shall here only play a role in as much as it is needed in the case of a discontinuously operated plant to attain the demanded operating pressure as quickly as possible after beginning a new processing run.

The calculation of the necessary pumping speed, needed to pump out always present leakage gas flows [5] and the process gas and vapour flows at a demanded operating pressure is based on the equation of state for ideal gases, whereby the vapours and the leakage flows which are present in any kind of process engineering plant need to be pumped out. The information which is available to the process engineer is usually stated by way of mass flows (kg h^{-1}) which need to be converted using the universal equation of state for ideal gases into volume flows, which are to be pumped at the desired process pressure. Based on this information and the conversion into volume flows, the magnitude of the pumping speed which needs to be provided by the vacuum pump to be procured or the vacuum system can be defined. Other properties of the process vapours like their corrosiveness, the tendency towards condensation or the formation of deposits, define the type or the material selection for the vacuum pump, respectively vacuum system which is to be installed.

References

1. Edelmann, C.H. (1978) *Vakuumphysik und -Technik*, Akademische Verlagsgesellschaft, Leipzig.
2. (a) Wutz, M., Adam, H., and Walcher, W. (1992) *Theorie und Praxis der Vakuumtechnik*, Vieweg-Verlag, Braunschweig/Wiesbaden. Jousten, K. (Hrsg.) (b) 2006) *Wutz, Handbuch der Vakuumtechnik*, Vieweg-Verlag.
3. O' Hanlon, J.F. (1989) *A User's Guide to Vacuum Technology*, John Wiley & Sons, Inc., New York.
4. (a) Voß, G. (2007) *Grundlagen und Prinzipien der Vakuumtechnik*, Haus der Technik, Essen. (b) Umrath, W. (2007) *Grundlagen der Vakuumtechnik*, Oerlikon Leybold Vacuum, Cologne.
5. *Vakuumtechnik für die Chemie*, Oerlikon Leybold Vacuum, Cologne.
6. Pupp, W. and Hartmann, H.K. (1991) *Vakuumtechnik, Grundlagen und Anwendungen*, Carl Hanser Verlag, München und Wien.

2
Condensation under Vacuum

Harald Grave

2.1
What Is Condensation?

The origin of the word 'condensation' is from the Latin 'condensare'. This means 'to seal up, to compact'. In this chapter, it stands for the compression and the congregation of molecules during the transition from the gaseous to the liquid state. For a physicist, this is the transition to a molecular motion with less energy. Therefore, condensation is combined with a release of energy. This energy is emitted as the condensation heat, which has to be dissipated. The reverse process is evaporation. Each liquid has a vapour pressure that increases with an increase in temperature; the relationship between the saturation temperature T_s and vapour pressure p_v of a substance is shown in the vapour pressure diagram. The vapour pressure curves for water and some solvents for a range from 1 mbar to 100 mbar are shown in Figure 2.1.

Here, the region to the right of the vapour pressure curve represents the gaseous state. To the left side of the curve, the substance is liquid or solid. In the condition described by the curve liquid and saturated vapour coexist in equilibrium. When the temperature of an unsaturated (superheated) vapour is reduced, condensation starts at the saturation temperature (dew point temperature) corresponding to the actual vapour pressure.

The heat of condensation – which is equal to the heat of evaporation – has to be drawn off during the condensation. The specific heat of condensation depends on temperature; for example, for water at $\vartheta = 25\,°C$ the heat of condensation is about 10% higher than at $\vartheta = 100\,°C$. For some substances, the heat of condensation is given in Table 2.1.

If the vapour is not saturated – according to the condition given by the vapour pressure diagram – but has a higher temperature, that is, the vapour is superheated, then the superheat also has to be drawn off.

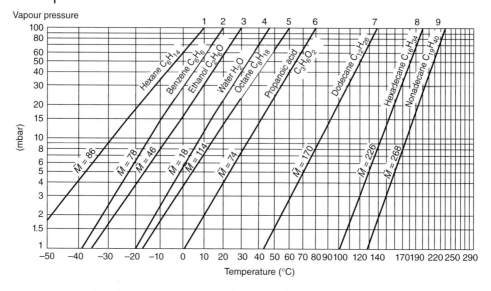

Figure 2.1 Vapour pressure diagram for some substances. (1) Hexane, (2) benzene, (3) ethanol, (4) water, (5) octane, (6) propanoic acid, (7) dodecane, (8) hexadecane, and (9) nonadecane.

Table 2.1 Molar mass M and specific condensation enthalpy r at 1013 mbar pressure for some substances.

Substance	Molar mass (kmol kg^{-1})	Heat of condensation (kJ kg^{-1})
Acetone	58.08	523
Benzene	78.11	394
Ethanol	46.07	846
Hexane	86.18	335
Octane	114.23	301
Water	18.02	2257

2.2
Condensation under Vacuum without Inert Gases

One can imagine that there is some liquid in a container. This liquid will be evaporated. A weightless, movable piston lies on the liquid surface, as represented in Figure 2.2. If the liquid is water and this is heated, as in Figure 2.3, then the water evaporates at 100 °C and the piston moves upward, so that the evaporation process can take place at constant pressure (1000 mbar) and at constant temperature (100 °C).

The space between piston and water surface is filled with steam at atmospheric pressure. Then, the heating is stopped, the piston is fixed at its actual position, and

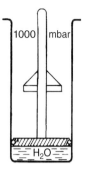

Figure 2.2 Container with weightless piston.

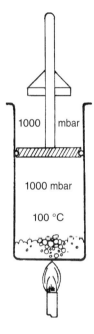

Figure 2.3 Boiling water and steam at atmospheric pressure.

the container is put into a cooling tank, as shown in Figure 2.4. Here, the complete equipment is cooled to 20 °C. The major part of the steam will condense and, if the equipment is sealed, a vacuum of 24 mbar can be measured. For the assumptions made, a vacuum is generated by condensation.

2.3
Condensation with Inert Gases

Starting from the condition in Figure 2.3, suppose that a certain amount of air (a noncondensable gas and in the following known as *inert gas*) was also included in the container before the cooling. With a total pressure of 1000 mbar, for example, a

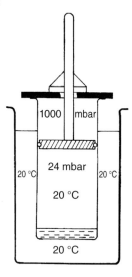

Figure 2.4 Steam and water in phase equilibrium.

water vapour partial pressure of 950 mbar is assumed. If one cools down to 20 °C with a constant volume, the mass of air remains the same and the partial pressure of air decreases from 50 to 40 mbar by the cooling from 100 to 20 °C. This partial pressure has to be added to the water vapour partial pressure of 24 mbar after water vapour condensation. So a vacuum of 64 mbar is created in this case. The remaining air–vapour mixture is saturated with water vapour, because the water vapour partial pressure is equal to the saturation pressure at the actual temperature of 20 °C. Instead of water, one can imagine any arbitrary liquids. Depending on the vapour pressure diagram for this media, other pressures will be reached.

Often condensable vapours are emitted continuously in vacuum processes, that is, at almost constant mass flow rates. In such a vapour flow, a certain inert gas fraction is normally contained, this can, for example, be a leakage air stream escaping into the vacuum equipment, which should, of course, be kept to a minimum. If this mixture flows along a suitable cooled surface at constant pressure, here, that is, under vacuum, then vapour will condense. The inert gas percentage, which was small at the beginning of the condensation process, will become larger with progressive condensation. Then, at the end of the condensation process, the remaining inert gas–vapour mixture must be removed. This is done via extraction by means of a suitable vacuum pump. The vapour portion, with which, at the end of the condensation surface, the inert gas is saturated, is naturally lesser with a reduction in temperature. Therefore, it is appropriate to extract at the coldest section of the condenser, that is, where the coolant is supplied. This is the reason why the coolant is usually taken in counterflow to the inert gas–vapour mixture that is to be condensed (Figure 2.5).

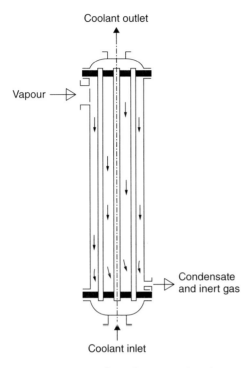

Figure 2.5 Counterflow of vapour and coolant in a surface condenser.

2.4
Saturated Inert Gas–Vapour Mixtures

An inert gas–vapour mixture saturated with condensing vapours has to be extracted independent of the condenser design. If there is only one kind of vapour, then the saturation quantity can be calculated easily with the equation:

$$\dot{m}_V = \dot{m}_I \cdot \frac{M_V}{M_I} \cdot \frac{p_V}{p_I} \tag{2.1}$$

M is the molecular weight, p is the partial pressure, and $\dot{m}$ the mass flow rate. The index 'I' stands for the inert gas and 'V' for vapour.

An example can best illustrate the influence the condensation conditions have on the specification of the vacuum pump.

Steam is to condense at a vacuum of 60 mbar. This corresponds to a saturated steam temperature of 36 °C. The steam to be condensed contains 10 kg h^{-1} of air. The condenser should be suitable to cool the air, together with the included steam down to 30 °C. The question is now, how much steam–air mixture has still to be extracted by a vacuum pump. The steam partial pressure of a saturated mixture with 30 °C is $p_v = 42.4$ mbar (= saturated steam pressure at 30 °C). The partial pressure of the inert gas is then the difference between the total pressure and the steam partial pressure.

$$p_i = 60 - 42.4 = 17.6\,\text{mbar}$$

Thus, the saturation quantity of steam is calculated with Equation 2.1:

$$\dot{m}_V = 10 \cdot \frac{18}{29} \cdot \frac{42.4}{17.6} = 15\,\text{kg}\,\text{h}^{-1}$$

This means, the vacuum pump must extract $15\,\text{kg}\,\text{h}^{-1}$ steam and the $10\,\text{kg}\,\text{h}^{-1}$ air from 60 mbar. It can be seen that the quantities of saturation steam are considerable in vacuum. The importance of good cooling of the outlet flow from a condenser is shown by the fact that for the above example a decrease in the outlet temperature from 30 to 25 °C would mean a reduction of the steam mass flow rate to only $7\,\text{kg}\,\text{h}^{-1}$.

However, the conditions are normally not as simple as in this example. Usually, mixtures of different kinds of vapours are condensed, and the individual components have different vapour pressure diagrams. The components can be perfectly soluble into each other, partly soluble, or insoluble.

2.5
Vapour–Liquid Equilibrium

For the example of a water and air mixture, as specified earlier, simple conditions also result, as the liquid phase comprises only one component. For a mixture with several components, the equilibrium must be determined, which means that the mole and/or mass fraction of the different components in both gas and liquid phases have to be calculated from appropriate sets of equations. For the gaseous phase, the sum of the partial pressures still results in the total pressure. This is described by the equation:

$$p_i = y_i p \tag{2.2}$$

where p_i is the partial pressure of the component i, y_i is the mole fraction of the component i in the gaseous phase, and p is the total pressure.

The correlation to the liquid phase for an ideal solution is given by the equation:

$$p_i = x_i p_{si} \tag{2.3}$$

where x_i is the mol fraction of the component i in the liquid phase and p_{si} is the vapour pressure of the pure component i.

From Equations 2.2 and 2.3, the above-mentioned Equation 2.1 can be derived easily for the system steam–air if one neglects the solubility of air in water and sets the mole fraction of water in the liquid phase $x_{H_2O} = 1$. If more than one component is contained in the liquid phase, then Equations 2.2 and 2.3 have to be solved iterative.

For many substances, the partial pressure over a diluted solution cannot be described with the ideal model of Equation 2.3. A method of resolution is the introduction of the activity coefficient γ. This coefficient depends on the substances that are concerned, their concentration, and the temperature:

$$p_i = x_i \gamma_i p_{si} \qquad (2.4)$$

For an ideal solution, $\gamma = 1$. Some substances have very large activity coefficients, which means that only very little condensate will form together with water. One can consider this fact by considering that these substances are not soluble in water, for example, this is true for oil.

It is possible for several liquid phases to exist, each reacting with a common vapour phase as there are no other liquid phases. An example makes this clearer.

Example

Water vapour and a substance X that is insoluble in water should condense. The molecular weight of X is $80\ \mathrm{kg\ kmol^{-1}}$, the pressure is 217 mbar, and the temperature is 36 °C. The vapour is also mixed with an air flow rate of $10\ \mathrm{kg\ h^{-1}}$.

The partial pressure of water vapour at 36 °C is $p_{H_2O} = 59.4$ mbar and the vapour pressure of X should be $p_A = 71.6$ mbar. Thus, the partial pressure of the inert gas is the remaining difference to the total pressure:

$$p_I = 217 - 59.4 - 71.6 = 86\ \mathrm{mbar}$$

With this data, the saturation mass flow rate of water vapour and for X can be calculated. One can do this with Equation 2.1, because only one liquid component has to be considered in each case:

$$\dot{m}_{H_2O} = 10 \cdot \frac{18}{29} \cdot \frac{59.4}{86} = 4.3\ \mathrm{kg\,h^{-1}}$$

$$\dot{m}_X = 10 \cdot \frac{80}{29} \cdot \frac{71.6}{86} = 23\ \mathrm{kg\,h^{-1}}$$

If less than $23\ \mathrm{kg\,h^{-1}}$ of substance X flow into the condenser, then there will be no liquid phase of X. In this case, X together with the air has to be regarded as not condensable. For example, an inert gas flow of $10\ \mathrm{kg\,h^{-1}}$ air + $10\ \mathrm{kg\,h^{-1}}$ X = $20\ \mathrm{kg\,h^{-1}}$, with a mean molecular weight of $42.6\ \mathrm{kg\,h^{-1}}$, would give a saturation vapour flow of

$$\dot{m}_{H_2O} = 20 \cdot \frac{18}{42.6} \cdot \frac{59.4}{(217 - 59.4)} = 3.2\ \mathrm{kg\,h^{-1}}$$

2.6 Types of Condensers

In principle, this can be simplified to the difference between direct-contact condensers and surface condensers. The first type is characterised by mixing the coolant with the vapour to be condensed; in the second type, surfaces separate coolant and vapour. Figure 2.6 shows different designs of direct-contact condensers.

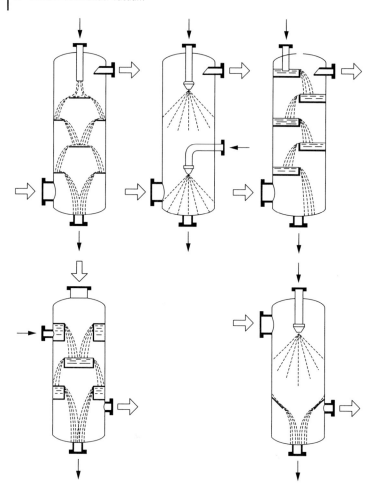

Figure 2.6 Different types of direct-contact condensers: upper row, counterflow; lower row, co-current flow.

Often these condensers have internal structures to create a better distribution of the coolant, so that the liquid surface is as large as possible. In spray condensers, nozzles are used for a good distribution of the coolant.

Direct-contact condensers are mostly built as counterflow condensers. There are, however, cases where one uses co-current flow, for example, if the inert gas–vapour mixture should not be under-cooled to too great a degree, then the components of the extracted product will form solids; here the co-current principle can offer advantages. Since with co-current condensers the vapours flow from top to bottom, with the force of gravity and in same direction as the coolant, the pressure losses are lesser and higher flow rates can be realised or smaller cross sections are sufficient for the duty. However, in co-current condensers, the mixture that has to be extracted is in contact with the heated coolant, the outlet temperature is higher than that of counterflow condensers; therefore, a larger mass flow rate has to be extracted. Sometimes, a co-current condenser is used for the main condensation to allow the overall dimensions to be as small as possible, and downstream of a counterflow condenser is used for the under-cooling of the exhaust mixture.

The advantages of a direct-contact condensation are the low purchase price and the best possible utilisation of the coolant. With direct-contact condensers, one can raise the temperature of the coolant nearly to the boiling temperature of the condensate without the need to build an extremely large condenser. This is because the heat transfer coefficients for such a direct heat transfer are very high. Therefore, a smaller cooling water flow rate is needed than with surface condensers. Furthermore, their insensitivity to fouling, high operating reliability, and easy maintenance are other advantages.

The biggest disadvantage of a mixing condenser is the fact that the vapour condensate is mixed with the coolant. This mixing is only acceptable when the condensate is harmless and is to be discarded after use. For all other applications, surface condensers have to be used.

With surface condensers, the condensation surface is usually formed by tubes, in which the coolant (normally cooling water) flows at speeds from 0.4 to 2 m s^{-1}. Vapour flows usually around the pipes, as with this construction sufficient cross sections for the large volume flow of the vapour can be realised. As an example of such a condenser, Figure 2.7 shows a tubular type condenser. For condensation

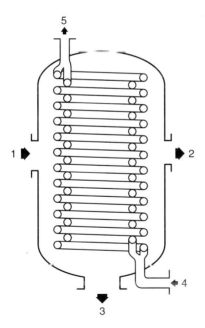

Figure 2.7 Tubular type condenser: (1) vapour inlet, (2) outlet for not condensed flow, (3) condensate outlet, (4) coolant inlet, and (5) coolant outlet.

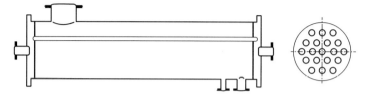

Figure 2.8 Tubular type surface condenser with fixed tubes: cooling water flows in the tubes; vapour flow in the shell room around the tubes.

surfaces up to certain square metres, this type is common in vacuum engineering. It allows both high cooling water velocities and large steam cross sections to be achieved simultaneously.

Figure 2.8 shows a horizontal tubular type condenser with fixed bank of tubes, where the condensation takes place around the pipes and the cooling water flows in the tubes. A disadvantage of this design is the inherent difficulty in cleaning the condensation area.

Figure 2.9 represents a surface condenser with fixed vertical tubes. Here, to allow for better cleaning, the vapour condenses inside the tubes, while the cooling water is led around the tubes. With this arrangement, there is poor cleaning on the cooling water side. With this design, the vapour enters the tubes with a very high velocity. This causes a good distribution and a thin condensate film. But care must be taken to ensure that sonic velocity is not reached at the inlet to the tubes. It should also be noted that how by means of baffles, the flow velocity of the coolant is increased, in order to increase the heat transfer.

Designs with floating head and extendable tube bundle are also possible (Figure 2.10). With this construction, the outer surface of the tubes can also be cleaned. This construction is often used in the chemical industry and at refineries. With this design, the possibility of good cleaning exists both on the vapour and on the coolant side. It should again be noted that how by using baffles, the flow velocity of the vapour is increased, in order to increase the heat transfer.

In addition to those mentioned, there are still a great number of further designs of surface condensers, for example, U-tube condensers, block condensers, and air condensers, all of which are used as vacuum condensers.

2.7
Heat Transfer and Condensation Temperature in a Surface Condenser

The necessary heat transfer area of a condenser can be calculated by the equation:

$$A = \frac{\dot{Q}}{k \cdot \Delta T_m} \tag{2.5}$$

where A is the heat transfer area (m^2), $\dot{Q}$ is the heat flow (W), K is the heat transfer coefficient (W m^{-2}K^{-1}), and ΔT_m is the average temperature difference (K).

Thus, the heat transfer coefficient and the average temperature difference must be known. For the condensation of pure saturated vapours, the process is only determined by the heat transfer in the condensate film and in the coolant. The gas phase has to be considered only if the vapour is superheated to a great degree. In the condensate film, the heat transfer is essentially influenced by the hydrodynamics of the film flow and by the thermal properties of the liquid. For the case of laminar flow of the condensate film, a calculation method was developed by Nußelt in 1916. Outside of its original application range, this method also gives good estimated values for the heat transfer.

For increasing film thickness, turbulence improves the heat transfer. In Figure 2.11, the flow patterns in vertical tubes are described for different inlet velocities of the vapour flow. For low velocities, first a laminar flow is formed. Then, after a transit phase, a turbulent film is developed. For high-inlet velocities, droplet entrainment can occur, which improves the heat transfer. After an annular flow, finally a plugged flow can occur.

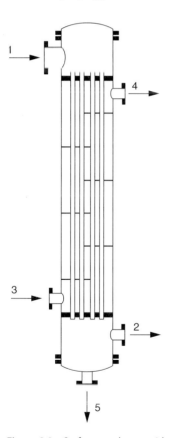

Figure 2.9 Surface condenser with fixed tubes. Condensation in the tubes, cooling water around the tubes. (1) Vapour inlet, (2) outlet for not condensed vapour and gas, (3) coolant inlet, (4) coolant outlet, and (5) condensate outlet.

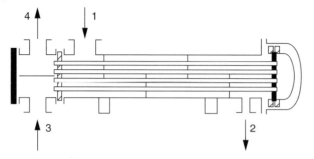

Figure 2.10 Floating head condenser. (1) Vapour inlet, (2) outlet for the not condensed vapour and condensate, (3) coolant inlet, and (4) coolant outlet.

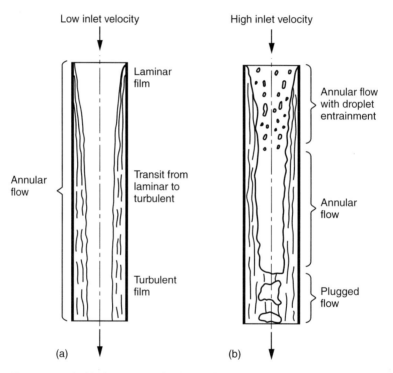

Figure 2.11 (a, b) Flow patterns by the condensation in vertical tubes. (Adapted from 'VDI-Wärmeatlas'.)

For horizontal tubes, it has to be considered that nonuniform circumferential distribution of the condensate film can occur. The heat transfer coefficient also depends on the circumference position (Figure 2.12).

If pure vapour, that is, only one component without inert gas, is condensed under vacuum, the whole condensation proceeds at a constant temperature. At the end of the condensation process, there is nothing left to be extracted. Such a

2.7 Heat Transfer and Condensation Temperature in a Surface Condenser

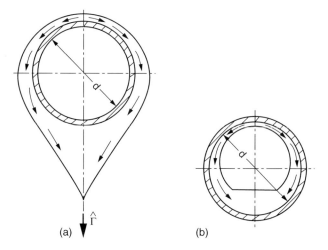

Figure 2.12 (a, b) Condensate film in horizontal tubes. (Adapted from 'VDI-Wärmeatlas'.)

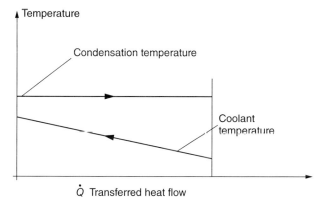

Figure 2.13 Temperature profile in a condenser for pure vapour condensation without inert gas.

condenser is an ideal sink for the vapour, which is condensed completely. At all places in the condensation space, the temperature is the same, that is, the saturation temperature of the vapour at the actual pressure. So the condenser is an ideal vacuum pump. It needs only coolants, which can absorb the heat of condensation.

Figure 2.13 shows the run of the condensation temperature and of the coolant temperature, plotted against the transferred heat flow. In this case, it also makes no difference whether the coolant is in counterflow or co-current flow to the vapour.

If mixtures of different vapours are condensed, or if an inert gas is involved, the efficiency of the condensation is degraded.

In the case of an inert gas, this gas has a partial pressure in the flow entering the condenser, depending on its mass fraction. This partial pressure can be calculated with Equation 2.1, which is transformed to

$$\frac{p_I}{p_D} = \frac{\dot{m}_I}{\dot{m}_D} \times \frac{M_D}{M_I}$$

And with the definition, the total pressure is the sum of the partial pressures:

$$p = p_I + p_D, \quad p_D = p - p_I$$

$$\frac{p_I}{p - p_I} = \frac{\dot{m}_I}{\dot{m}_D} \cdot \frac{M_D}{M_I}$$

It follows:

$$p_I = p \frac{\frac{\dot{m}_I M_D}{\dot{m}_D M_I}}{1 + \frac{\dot{m}_I}{\dot{m}_D} \cdot \frac{M_D}{M_I}} \quad (2.6)$$

The partial pressure of the vapour is reduced, in relation to the total pressure, by this inert gas partial pressure from Equation 2.6. This means, the condensation of the vapour does not start at the saturation temperature relating to the total pressure, but condensation starts at the lower saturation temperature of this reduced partial vapour pressure. During the condensation process, the inert gas fraction increases, as parts of the vapour are condensed. So the condensation temperature decreases more and more. The development of the temperature is described by Figure 2.14.

A comparison of Figure 2.14 with Figure 2.13 shows temperature differences that are substantially more unfavourable. Owing to the inert gas fraction, the condensation temperature is already decreased at the condenser inlet. It continues to drop during the process of condensation and reaches its lowest value at the end of the condenser, as you find the highest fraction of the inert gas in the yet not condensed mixture. As mentioned earlier, the aim is to get a vapour fraction as small as possible at the condenser outlet, that is, to cool down as far as possible, which is limited by the available coolant temperature.

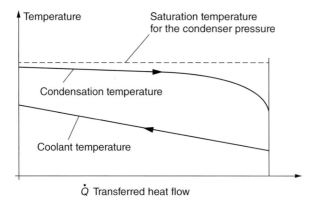

Figure 2.14 Temperature profile in a condenser with inert gas.

The temperature differences ΔT at each point in the condenser are smaller than for the condensation of inert gas free pure vapour. Particularly at the end of the condenser (after the main amount of condensation, heat is dissipated), the condensation temperature drops to very low values. It can easily be seen that in these cases, it is very important to run the coolant in counterflow to the condensing vapour. Sometimes, such condensers are constructed with a special under-cooling section where the remaining inert gas–vapour mixture is forced to flow to the coldest place of the condenser. Installations of the most diverse forms are made, in order to achieve this purpose. However, in such installations, dead corners can form when the flow is not directed properly. There the inert gas collects and blocks the vapour away from the heat exchange surface, so that parts of the surface remain unused.

Thus, by the presence of inert gases, not only does the average temperature difference ΔT_m become smaller but also the heat transfer degrades, since the heat transfer resistance on the condensation side increases. The cold walls are a sink for vapours. If vapour condenses, the inert gas remains and forms a layer at the heat exchange surface. In order to get to the cold wall, the vapour must diffuse through this inert gas layer, thus heat transfer is made more difficult. This effect is more pronounced: the higher the inert gas fraction is or becomes during the condensation. Therefore, the heat transfer is worst at the condenser outlet.

The inert gas layer at the heat exchange surface can be flushed away by a high flow rate in the condensation area, this increases the heat transfer. Similarly, the heat transfer is corrupted by a decrease in the flow velocity. During the condensation, however, the volume flow rate is reduced, and with the same flow area the velocity is also reduced. At the same time, the condensation temperature drops. These are all factors that cause the condensation conditions to become more unfavourable near the condenser outlet.

The heat transfer coefficients of vacuum condensers depend on many different factors, particularly on the inert gas content and the flow rate in the condensation area. Relatively high heat transfer coefficients exist, for example, in turbine condensers, where sometimes enormous quantities of turbine exhaust steam are condensed under vacuum. The heat transfer coefficients depend on the condenser design and the inert gas content. Values from 2500 to 5000 W $(m^2K)^{-1}$ are common. In this case, pure water vapour has to be condensed, so that on the vapour side no fouling is to be taken into consideration. On the water side, the fouling depends on the cooling water quality, and therefore appropriate fouling factors have to be considered.

In condensers, in which vapour mixtures are condensed, for example, hydrocarbon vapours from distillation columns in oil refineries, heat transfer coefficients between 200 and 2500 W $(m^2K)^{-1}$ arise. Here, an accurate computation is very difficult, because in such a condenser in different places different components can condense in different places with changing inert gas fractions, changing flow rate and direction of flow, and at changing condensation temperatures. The uncertainty of the computation is often compensated by the fact that on the vapour and on the water side, intense fouling can occur. Hence, refineries indicate fouling factors of about $0.0002\,m^2 K\,W^{-1}$ on the water side and of about $0.0004\,m^2 K\,W^{-1}$ on the vapour side. This corresponds to heat transfer coefficients of 2500 and $5000\,W\,(m^2 K)^{-1}$ and means a double or triple size of the condensation surface in comparison to a completely clean condenser. Thus, an error in the computation of the heat transmission coefficients on the vapour side only affects the cleaning cycle time. However, in its new condition, the condenser always has a sufficient reserve for a perfect function. Therefore, when purchasing a condensation plant, the offers from different suppliers have to be compared, regarding the calculated heat exchange area, since this naturally determines the price.

2.8
Vacuum Control in Condensers

In a condensation plant, the vacuum depends on several factors: the vapour mass flow rate, the inert gas fraction, the fouling of the condenser, the composition of the vapour (if several components are present), the coolant mass flow rate, and the coolant temperature. These are the inputs, which can vary in an existing plant. Fixed values are, for example, the heat exchange surface, the design of the condenser, and the size of the vacuum pump. These things cannot be changed quickly. If control is needed, for example, to keep the vacuum constant, one should consider first, which variables will change and affect the vacuum. Then, one must consider how these parameters can be kept constant or, if this is not possible, one has to look for variables that can be changed to cancel out the fluctuations. In most cases, the coolant flow rate is the suitable variable. In general, one can affect the vacuum by changing the condensation conditions. If possible the best option is to change the variable that has the strongest influence on the condensation conditions. Also the addition of inert gas, for example, ballast air, changes the condensation conditions. However, this can only be varied within the range in which the vacuum pump is able to extract this additional amount of air.

2.9
Installation of Condensers

A very common layout for the assembly of condensers is barometric installation, the method in which no pump is needed to extract the liquid from under vacuum. The condenser is erected at such a height above the liquid drain level that the pressure difference between the vacuum inside the condenser and the atmospheric pressure is compensated by a liquid column. The height of this barometric leg must also overcome the flow resistance. In Figure 2.15, such a barometric installation is shown for a unit employing surface condensers. The height between the condensate outlet at the condenser and the liquid surface level in the collecting tank can be calculated by the necessary pressure difference and the specific weight of the

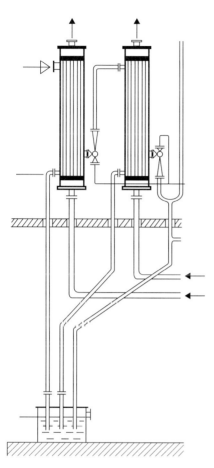

Figure 2.15 Barometric installation of surface condensers.

liquid. For water, a height of 11 m is suitable for condenser pressures of 100 mbar or less. It is important that the drain pipes have a suitable gradient and are not a combination of vertical and horizontal sections.

If there is not enough height available for a barometric installation, a pump has to be installed. Usually, such a pump is controlled by the liquid level above, so that there is always a liquid seal to prevent the inflow of air. Alternatively, a nonreturn valve must be mounted at the discharge side of the pump, but then cavitation has to be anticipated, if the drain pipe from the condenser is completely emptied by the pump. Collecting vessels are also used to collect the condensate under vacuum, discharging as required. For continuous operation, two or more alternating vessels are needed, and the ventilation and evacuation have to be considered.

When several condensers with different pressures have to be drained, a 'semi-barometric' installation is possible with a closed collecting tank instead of an open tank. In this tank, an interim vacuum exists, so that a given height is used and only one drain pump is needed. Such a system is illustrated in Figure 2.16.

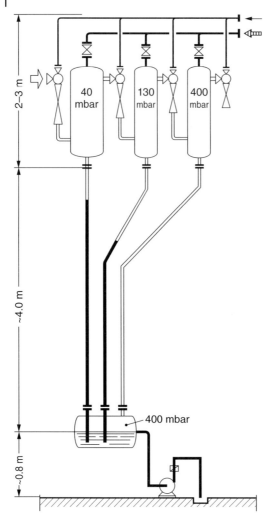

Figure 2.16 Semi-barometric installation with collecting tank and drain pump.

2.10
Special Condenser Types

If vapours have to be condensed in a vacuum process at a pressure below the triple point, then this can be done without previous increase in pressure, but only as solids. For this desublimation of vapours, one uses changeover condensers. The first condenser will always be loaded with the product, while the product in the second is melted. For this procedure, a suitable coolant is naturally required, usually brine or evaporating refrigerant of a refrigeration process. For example, during a freeze-drying process such ice condensers are used, because the water vapour has to be removed from the frozen product at a temperature of approximately −20 °C and is condensed in the form of ice at approximately 1 mbar.

As mentioned earlier, direct-contact condensers have some advantages against surface condensers. However, the main disadvantage is the mixing of vapour condensate with the coolant. If pure steam is condensed with water as coolant, this may not be a problem. But if it is not permitted to mix cooling water and condensate and in addition there is a fouling problem, then it is difficult to find an adequate condensation system. On the one hand, direct-contact condensers

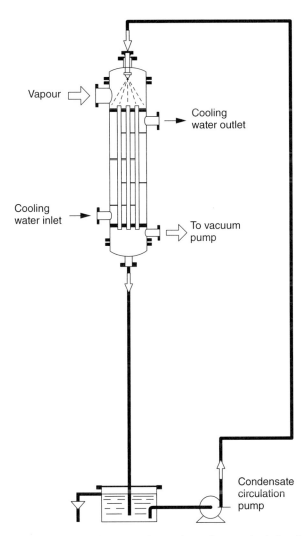

Figure 2.17 Condensation plant with condensate circulation. Superheat is dissipated to prevent buildup of solid incrustations at the inlet part of the heat exchanger tubes.

should be chosen because of the fouling problem, but on the other, surface condensers are preferable because of the contamination of the coolant. This problem can be solved by using both principles. A direct-contact condenser is used, but the vapour condensate itself is used as coolant. The condensate is pumped in a circuit and is cooled in a surface heat exchanger with the normal cooling water. Often this condensate is also a good solvent for the removal of any fouling or incrustation.

The insensitivity of direct-contact condensers to fouling is a result of the high amount of liquid that is flowing along the walls and internals. This feature can even be improved by a suitable design. A direct-contact condenser should be designed in such a way that there are no internal dry spots, where solid incrustations can accumulate.

A similar improvement is also possible for surface condensers. If the condensation takes place inside the tubes, usually there is a complete condensate film at the lower part of the tubes, which prevents solids buildup. However, at the upper part, near the vapour inflow, the superheat has to be dissipated and the tubes can have parts with dry surfaces. Therefore, these condensers are fouled mainly in this zone. To prevent this, it is possible to inject some recirculated condensate onto the upper side of the tube sheet. Figure 2.17 shows such an installation that can often prevent the fouling of the tube surface.

Further Reading

Jousten, K. (ed) (2008) *Handbook of Vacuum Technology*, Wiley-VCH Verlag GmbH, Weinheim, pp. 353–374.

VDI-Gesellschaft Verfahrenstechnik (1997) *VDI-Wärmeatlas*, 8 Auf, Springer-Verlag, Heidelberg.

Power, R.B. (1993) *Steam Jet Ejectors for the Process Industries*, McGraw-Hill, pp. 105–150.

3
Liquid Ring Vacuum Pumps in Industrial Process Applications
Pierre Hähre

3.1
Design and Functional Principle of Liquid Ring Vacuum Pumps

3.1.1
Functional Principle

The main components of a liquid ring vacuum pumps, a special version of liquid ring compressors are a cylindrical casing with, for example circular cross section and casing cover discs as well as an impeller with radial blades.

If one fills this casing up to about the shaft height with an appropriate liquid and moves the impeller in rotary motion, then a uniform liquid ring is formed with a sufficiently high number of revolutions at the casing wall because of the centrifugal forces acting on the fluids. The gaseous medium left in the casing is distributed in the hub area around the impeller because of its lower density (Figure 3.1).

If the rotation axis of the impeller is displaced radially in the casing, this is rather without influence on the formation of the liquid ring. However, the space available for the gaseous medium between the liquid surface and hub assumes a corrugated form. This corrugated space is subdivided several times by the blades in a circumferential direction.

If one selects the blade base between the two blades (moved reference system) as a point of reference, then one can observe with the rotation of the impeller that, according to Figure 3.2, the liquid surface first affects the impeller hub, departing from the observation point in the further course of the rotary motion, in order then to approach it again. One recognises a piston motion, by which the space between two blades and the liquid surface is first enlarged and then reduced again (Figure 3.2).

If one now places an opening in the area of the enlargement of the space enclosed by the impeller and liquid ring, then this volume can be filled by a medium to be fed of lesser density, usually gas. An analogous opening in the area of the decreasing chamber volume makes the discharge of the lighter fluid possible (Figure 3.3).

Vacuum Technology in the Chemical Industry, First Edition. Edited by Wolfgang Jorisch.
© 2015 Wiley-VCH Verlag GmbH & Co. KGaA. Published 2015 by Wiley-VCH Verlag GmbH & Co. KGaA.

Figure 3.1 Formation of a liquid ring.

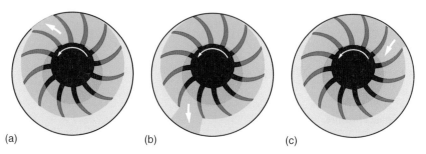

Figure 3.2 Realisation of the displacement principle with liquid ring compressors. (a) Liquid ring touch the hub, (b) liquid ring remove from the hub and (c) liquid ring get back to the hub.

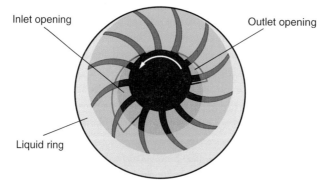

Figure 3.3 Working openings.

In contrast to other displacement compressors, liquid ring machines have a fluid 'piston'. This is subject to a constant motion and deformation depending on effective forces. Potential differences are rapidly equalised by the motion of the fluid inside the piston. If the ring liquid plunges into the blade area, then the pressure of the enclosed medium is increased.

The counterforce necessary, therefore, must be applied by the liquid ring. The energy necessary for the delivery and compression is therefore almost exclusively

3.1 Design and Functional Principle of Liquid Ring Vacuum Pumps | 37

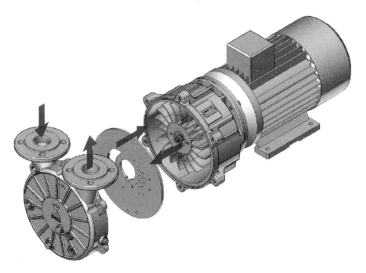

Figure 3.4 Exploded view of a single-sided pressurised liquid ring compressor.

extracted from the liquid ring. The kinetic (motion-) energy supplied to the liquid ring at the beginning of the suction process via the blades must be converted during the delivery and compression process into potential (pressure-) energy.

3.1.2
Design Details

3.1.2.1 The Medium Conveyance
Regarding the supply and removal of the medium to be pumped by the front surfaces of the respective compressor stage, one distinguishes single-sided and double-sided pressurised machines.

Figure 3.4 shows an exploded view of a single-sided pressurised machine. The gas exchange occurs via only *one* inter casing with suction and pressure port. This principle is realised in the closed coupled design of all single-stage liquid ring vacuum pumps (LRVPs).

In Figure 3.5, a double-sided pressurised machine is described. It possesses two inter casings with suction and pressure ports, so that an exchange of the medium to be compressed can take place via both front faces of the compressor stage. This type of arrangement is necessary if large suction capacities are to be realised with comparatively long impellers.

But to some extent, one also guides the medium to be pumped from one to the other front face of the compressor stage. In addition, one inter casing with a suction port and another inter casing with a pressure port are needed.

3.1.2.2 Design of the Pressure Ports
Until a blade cell has reached the end of the intake opening during a revolution, the compression of the enclosed gas cannot begin. From Figure 3.3, it is to be inferred

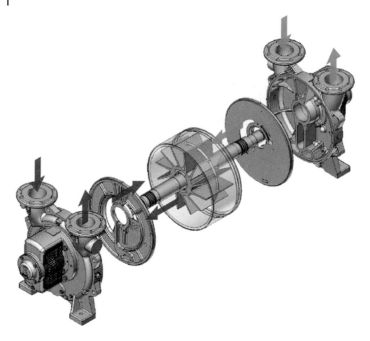

Figure 3.5 Exploded view of a double-sided pressurised liquid ring compressor.

that based on the volume reduction, the pressure excising in the blade cells varies with the angle of rotation. The gaseous medium is compressed up to reaching the pressure port.

If the counterpressure at the fixed pressure port is less than the pressure achieved by the volume reduction between the blades, then the compressed medium to be pumped is released via the pressure port.

Similar to greater counterpressure, the gas is released first at the pressure port into the blade area and it pushes out again in the further course of the rotary motion. The losses arising as a consequence of the restriction at the pressure port must be counterbalanced by an increased energy input. Consequently, with a pressure port corresponding to Figure 3.6, the liquid ring machine works optimally only with a specific pressure ratio.

The so-called variable pressure ports can be used for the reduction of the restrictor losses. These are generally drillings and slots covered by mobile plates and balls, which open or close depending on the difference of pressure at the valve. The valves are not opened until the pressure, which depends on the angle of rotation and suction pressure in the blade area, is greater than the counterpressure. One can thereby substantially reduce the restrictor losses.

However, the power requirement of the compression to be made available by the ring liquid is also reduced by the reduction of the restrictor losses. The liquid ring is decelerated less, so that the cross section passed through is reduced. Thus, the area available for the gas is enlarged, so that an increase in the suction capacity is to be denoted.

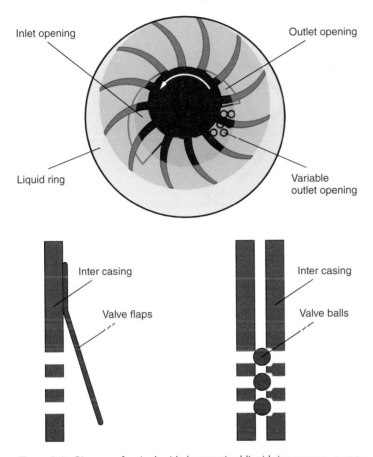

Figure 3.6 Diagram of a single-sided pressurised liquid ring vacuum pump.

The arrangement of the valves such that the surface is not covered leads to a pulsating gas discharge. A constant movement of the valves results from this. In comparison to the flaps, the valve bodies are subject to greater wear by the constant impact load. In the case of sufficient decrease in the diameter of the balls, the balls fall through the drillings. The result is a reduction in the suction capacity combined with an increase in the power consumption.

3.1.2.3 Situation of the Working Openings

For reasons of economy and also for reasons that are historically developed, primary vacuum pumps with inter casings have prevailed in the industry up to now. Deformations of the impeller and inter casing rotationally symmetrical to the shaft centre are conceivable for the enlargement of the working openings. Conical and cylindrical hubs are already completed constructions (Figure 3.7). The possibilities of the medium conveyance discussed in the earlier details as well as the design of the pressure ports are also transferable to the employment of these control bodies.

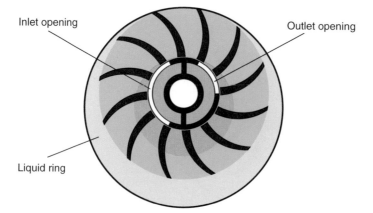

Figure 3.7 Diagram of a liquid ring compressor with port cylinder.

3.2
Operating Behaviour and Design of Liquid Ring Vacuum Pumps

The following requirements for the operating liquid result from the principle of operation and the design:

- Transfer of energy from the impeller to the medium to be compressed
- Sealing of the cells in radial direction
- Sealing of the gap between the impeller and the casing components
- Cooling and lubricating of the shaft sealing.

The ring liquid has to fulfil further requirements depending on the field of application. These might be, for example,

- absorption of compression and frictional heat;
- absorption of condensation and reaction heat;
- absorption of gaseous impurities from the medium to be pumped;
- absorption of incoming particles.

The delivery characteristics of LRVPs are affected by the properties of the ring liquid and of the medium to be pumped.

3.2.1
Hydraulics

3.2.1.1 Power Consumption
A part of the power consumption arises from the expenditures for the isothermal (running at constant temperature) compression of non-condensable gas

$$P_{\text{Gas}} = p_S \cdot \dot{V}_{\text{Gas}} \cdot \ln\left[\frac{p_G}{p_S}\right] \tag{3.1}$$

and for the pressure increase in the liquid pumped with it

$$P_F = (p_G - p_S) \cdot \dot{V}_F \tag{3.2}$$

Furthermore, friction losses in the liquid ring and at the walls as a consequence of the rotation of the liquid must be counterbalanced. By simplifications of the impulse and energy conservation laws of fluid mechanics, the following statement can be given therefore as

$$P_V = \frac{\rho_F \cdot \dot{V}_{FR}}{2} \cdot \zeta \cdot v^2 \tag{3.3}$$

Contrary to the expenditures for compression (Eq. (3.1)) and pressure increase (Eq. (3.2)), power dissipation loss depends on the material data of the ring liquid. Eq. (3.3) refers to a proportional impact of the density ρ_F on the power dissipation loss. The impact of other parameters such as the flow regime, the geometry and the viscosity is summarised in the pressure loss coefficient ζ.

Since a sufficient statement of the connections would go beyond the scope and purpose of these details, only the qualitative statement to be met here is that the use of highly viscous operating liquids increases the power dissipation loss. In case of doubt, an interpretation by the manufacturer is required.

3.2.1.2 Suction Volume

Depending on the functional principle, LRVPs are assigned to piston compressors. However, due to the flow behaviour of the ring liquid, almost the entire power requirement for the compression (Eq. (3.1)), the pressure increase (Eq. (3.2)) and overcoming the friction losses (Eq. (3.3)) are extracted from the liquid ring. The liquid ring is decelerated once again by the power consumption after entry of the energy via the blades:

$$P_{Gas} + P_F + P_V = \frac{\rho_F \cdot \dot{V}_{FR}}{2}(v_1^2 - v^2) \tag{3.4}$$

Due to the constancy of the circulating liquid quantity, the deceleration leads to an extension of the liquid ring to the impeller hub:

$$\dot{V}_{FR} = v \cdot A = v_1 \cdot A_1 \tag{3.5}$$

One can infer from both of these equations that with increasing power consumption, the liquid ring decelerates more strongly (Eq. (3.4)) and thus the suction chamber is reduced (Eq. (3.5)). It is to be recognised, for example that highly viscous liquids not only raise the power consumption of the vacuum pump but also impair the suction capacity.

3.2.2
Thermodynamics

The working chamber available for the compression is limited by the impeller hub; inter casings or casings; and the liquid ring. Since its size depends only on the hydraulics of the LRVP in the respective operating point, the suction capacity resulting therefrom is to be denoted in the following as the hydraulic suction capacity $\dot{V}_{Hydr}$.

3.2.2.1 Condensable Components in the Medium to Be Pumped

If the condensable components are contained in the pumped flow, then these, depending on the material properties, condense during contact with the operating liquid. The chamber becoming open due to the condensation is filled up by gas flowing in, so that an increased suction capacity is to be denoted as

$$\dot{V}_S = \dot{V}_{Hydr} + \dot{V}_D \qquad (3.6)$$

Figure 3.8 illustrates the vapour portion in saturated humid air at a temperature of 25 °C. The water vapour pressure is approximately 32 mbar, so that with a suction pressure of 32 mbar only pure water vapour is present.

In the assumed case of the position of equilibrium and in consideration of the Dalton law: $p_S \cdot \dot{V}_D = p_D \cdot \dot{V}_S$, the suction capacity results from Eq. (3.6):

$$\dot{V}_S = \frac{\dot{V}_{Hydr}}{1 - \frac{p_D}{p_S}} \qquad (3.7)$$

3.2.2.2 Temperature of the Medium to Be Pumped

On the basis of the intensive phase contact, it can be assumed with sufficient approximation that the pumped flow entering into the LRVP is warmed up or cooled down to the liquid temperature in the suction area. Thus, the following

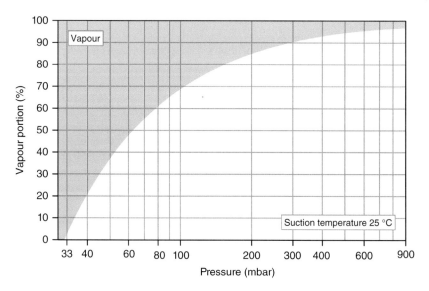

Figure 3.8 Vapour portion in saturated humid air at a temperature of 25 °C.

applies for the suction capacity:

$$\dot{V}_S = \dot{V}_{Hydr} \frac{T_S}{T_F} \tag{3.8}$$

3.2.2.3 Operating Liquid

Because of the functional principle of LRVPs, an intensive phase contact exists between the operating liquid and the medium to be compressed. Therefore, the available working chamber V_{Hydr} is partially filled by the vapour of the ring liquid V_F in addition to the medium to be pumped. Thus, the suction capacity is reduced by the vapour of the ring liquid:

$$\dot{V}_S = \dot{V}_{Hydr} - \dot{V}_F \tag{3.9}$$

If one proceeds from the position of the equilibrium condition, the partial vapour pressure and the volume enclosed by the vapour of the ring liquid are determined by the temperature of the ring liquid.

Figure 3.9 illustrates the volume damage at a water ring temperature of 25 °C. With declining pressure, the volume portion of the water vapour produced is increased. If the suction pressure achieves the equilibrium pressure of approximately 32 mbar, the vapour fills the entire chamber so that no further air can be absorbed.

Considering the Dalton law, $p_S \cdot \dot{V}_F = p_F \cdot \dot{V}_{Hydr}$ results from Eq. (3.9) for the suction capacity,

$$\dot{V}_S = \dot{V}_{Hydr} \cdot \left[1 - \frac{p_F(T_F)}{p_S} \right] \tag{3.10}$$

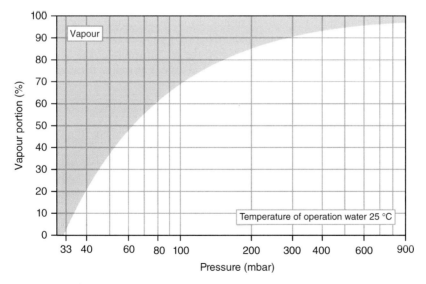

Figure 3.9 Volume damage in a liquid ring compressor by the vapour of the operating liquid at a water temperature of 25 °C.

If one combines Eqs. (3.7), (3.9) and (3.10), a model for describing the elucidated procedures is obtained:

$$\dot{V}_S = \dot{V}_{Hydr} \cdot \frac{p_S - p_F(T_F)}{p_S - p_D} \cdot \frac{T_S}{T_F} \qquad (3.11)$$

With the application of these generally accepted equations, it is to be considered that contrary to the common literature, the operating liquid temperature corresponds here to the temperature of the liquid ring and not to that of the supplied liquid.

The supplied operating liquid is warmed up in the vacuum pump by the absorption of the compression energy, the condensation heat and the heat of the discharged gas.

One can assume in a simplified manner that the gas and liquid temperature are identical at the pressure-side connection. If the suction medium is significantly colder than the ring liquid, then the cooling down of the ring liquid can also occur here. The energy balance around the LRVP reads

$$\dot{Q}_G + \dot{Q}_D + \dot{Q}_V = Q_F \qquad (3.12)$$

with

$$\dot{Q}_G = \dot{m}_G \cdot c_{PG} \cdot (t_S - t_F) \qquad \text{Heat input by inert gas}$$
$$\dot{Q}_D = \dot{m}_D^S \cdot h(t_S, p_S) - \dot{m}_D \cdot h(t_F, p_U) \qquad \text{Heat input by vapour}$$
$$\dot{Q}_V = P_W \qquad \text{Heat input by compression}$$
$$Q_F = \dot{m}_F \cdot c_{PF} \cdot (t_F - t_F^0) \qquad \text{Heat input by the ring liquid}$$

(3.13)

With the help of these equations, one can compute the actual temperature in the liquid ring on the basis of the hydraulically necessary liquid requirement.

A regulation of the operating liquid temperature should take place in principle on the basis of the ring liquid escaping on the pressure side of the compressor stage, since otherwise energy inputs by gas and vapour as well as the variability of the amount of liquid remain unconsidered.

3.2.2.4 Catalogue Values

However, one finds no hydraulic suction capacity in the catalogue. The characteristics described here apply mostly to the compression of dry air and of an operating water temperature of approximately 18 °C (15 °C + 3 K heating). Thus, in these values, the modification of the suction capacity by the vapour of the ring liquid and by the temperatures of gas and ring liquid are already considered.

However, if the operating conditions diverge from those in the catalogue, then first the hydraulic and hence the actual suction capacity can be determined from the characteristics with aid of Eq. (3.11). Thus, taking into consideration, the conditions specified in the catalogue yields

$$p_F(T_F = 291 \text{ K}) = 21 \text{ mbar} \qquad T_S = 293 \text{ K}$$
$$p_D = 0 \text{ mbar} \qquad T_F = 291 \text{ K}$$

with a suction pressure p_S and the related suction capacity $\dot{V}_K$ for the hydraulic

$$\dot{V}_{Hydr} = \dot{V}_K \cdot \frac{p_S}{p_S - 21\,\text{mbar}} \cdot \frac{291\,\text{K}}{293\,\text{K}} \tag{3.14}$$

and thus for the actual suction capacity:

$$\dot{V}_S = \dot{V}_K \cdot \frac{p_S}{p_S - 21\,\text{mbar}} \cdot \frac{291\,\text{K}}{293\,\text{K}} \cdot \frac{p_S - p_F(T_F)}{p_S - p_D} \cdot \frac{T_S}{T_F} \tag{3.15}$$

3.2.3
Counterpressure, Air Pressure

With increasing altitude above sea level, the ambient air pressure declines. For this, the following applies

$$p_U^h = p_U^0 \cdot e^{\left[-\frac{g}{R_K \cdot T} \cdot h\right]} \tag{3.16}$$

The reduction of the acceleration due to gravity with the increasing altitude is negligibly small. In the following, an ambient pressure $p_U^0 = 1013$ mbar, a temperature of 20 °C ($T = 293.15$ K) and an acceleration due to gravity of $g = 9.81$ ms^{-2} are assumed. The specific gas constant R_K has a value of 287 kJ/kg K.

From the reduction of the air pressure, a smaller power consumption results in the case of compression against ambient pressure with LRVPs with variable pressure slot. This consists of the compression achievement

$$P_{\text{isoth}} = p_S \cdot \dot{V}_S \cdot \ln\left(\frac{p_U}{p_S}\right) \tag{3.17}$$

and the power losses essentially independent of the compression pressure:

$$P_V = P_{\text{isoth}} \cdot \left[\frac{1}{\eta_{\text{isoth}}} - 1\right] \tag{3.18}$$

The degree of efficiency necessary for this can be determined from the characteristics for the displacement behaviour and the power consumption described in the catalogue:

$$\eta_{\text{isoth}} = \frac{P_W}{p_S \cdot \dot{V}_S \cdot \ln\left(\frac{p_U}{p_S}\right)} \tag{3.19}$$

thus resulting from Eqs. (3.15) and (3.16)

$$\frac{p^h}{p^0} = 1 + \eta_{\text{isoth}} \cdot \left[\frac{p^h_{\text{isoth}}}{p^0_{\text{isoth}}} - 1\right] \tag{3.20}$$

In order to prevent overloading of the installed electric motors, one should be geared to the maximum power consumption. Since the modification of the power requirement is essentially determined by the work done on compression, the maximum of the compression achievement also represents the maximum of the power consumption. This can be derived from Eq. (3.21):

$$P_{\text{th,max}} = p_U \cdot e^{(-1)} \cdot \dot{V}_S(p_S = p_U \cdot e^{(-1)}) \tag{3.21}$$

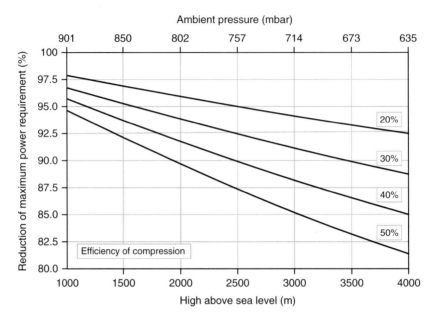

Figure 3.10 Reduction of the maximum power requirement for liquid ring vacuum pumps with variable outlet slots in open systems depending on the altitude.

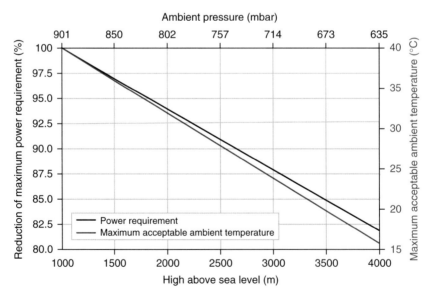

Figure 3.11 Reduction of the motor power depending on the altitude.

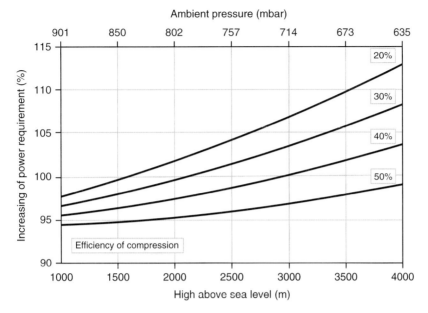

Figure 3.12 Necessary motor power requirements for open systems depending on the height above sea level for the operation employment of liquid ring vacuum pumps with variable outlet slots.

thus resulting from Eqs. (3.18) and (3.19) for the power requirement in use in higher altitudes:

$$\frac{P^h_{max}}{P^0_{max}} = 1 + \eta_{isoth} \cdot \left[\frac{P^h_U}{P^0_U} - 1\right] \tag{3.22}$$

Furthermore, it is, however, to be considered with the layout design that the electric motors installed for the provision of the shaft power are more poorly cooled due to the lower air density, whereby their load capacity is reduced. For this purpose, different engine manufacturers give the standard values described in Figure 3.11 for output and ambient temperature.

The summary of the statements of Figures 3.10 and 3.11 leads to the motor design described in Figure 3.12, which can be considered in the employment of LRVPs with variable pressure slots in higher altitudes.

3.2.4
Design Data

In reference to the remarks made earlier, the following data, required for the examination of the suitability and design of the LRVPs, are necessary:

1) Medium to be compressed
   a. Temperature
   b. Negative pressure to be produced

c. Quantity to be absorbed
d. Material composition
   i. Collection of condensable and non-condensable components
   ii. Impurities due to liquid droplets and particles regarding possible abrasion and corrosion problems
2) Operating liquid
   a. Temperature
   b. Material composition
      i. Mass fractions at liquid mixtures
      ii. Dissolved gas and solids.

The most exact material compositions possible are crucial here, since the operating behaviour depends essentially on the material properties of both phases.

3) Ambient conditions
   a. Temperature
   b. Pressure
   c. Humidity
   d. Installation site (inside/outside).

3.3
Vibration and Noise Emission with Liquid Ring Vacuum Pumps

LRVPs, similar to all rotary piston displacement compressors, are characterized by small vibrations and low-pulsation delivery. Nevertheless, if the operating performance does not meet the requirements of the customer, the vibration and sound emissions can be reduced by more or less extensive measures. On behalf of the most cost-effective solution possible, it requires well-founded knowledge of the procedures in the vacuum pump.

Sound consists of mechanical vibrations and waves of an elastic medium. It is produced by a mechanical stimulation of the elastic medium. According to the carrier media, one distinguishes gas, liquid or impact sound. Sound can be perceived by the human ear in the frequency range of 16–20 000 Hz.

3.3.1
Vibration Stimulation by Imbalance of Rotary Solids

The microscopic and macroscopic inhomogeneity of all solids causes an unequal distribution of the masses. If the solid accelerates, then different forces affect it depending on the mass distribution. If the body is to follow a certain movement, then resulting forces deviating by the direction of motion must be compensated by reactive forces.

If it concerns a rotation in the case of the movement of the body, then its inhomogeneity leads to a so-called imbalance. Depending on the number of revolutions $\vec{n} = \vec{\omega}/2\pi$ and the mass distribution asymmetrical to the rotation axis

Δm_R, the resulting forces $\vec{F}_R$ perpendicular to the rotation axis act over the entire lengths of the component that rotate with the solids:

$$\vec{F}_R = \vec{a}_R \cdot \Delta m_R = (\vec{\omega} \times \vec{r}) \cdot \Delta m_R \tag{3.23}$$

The counterforce necessary for the compensation is to be applied by the carrying components of the aggregate. If the compensation takes place statically, then the forces proceeding from the rotating component are compensated by an acceleration of the carrier construction:

$$\vec{F}_T = \vec{F}_R \qquad \vec{F}_T = \vec{a}_T \cdot m_T \tag{3.24}$$

With the increasing mass of the carrier construction, the acceleration acting on the aggregate is reduced. Here one speaks of the so-called inertia of the carrier elements.

Since the constantly changing forces significantly stress the carrier system, in the interest of as long an operating time as possible, the rotating machine elements are balanced along the rotation axis, that is materials are applied or removed, in order to adjust the differences of the mass distribution. Yet, production and economic limits are set to the balance quality, so that a rest unbalance always continues to exist.

The direction of the resulting force periodically changing due to the rotation with constant rotational speed stimulates vibration in all the components connected with the aggregate.

3.3.2
Vibration Stimulation by Pulsation

The LRVPs, similar to all displacement compressors, feed the medium to be compressed with the help of a limited number of chambers, which are periodically filled and drained. Pulsations occur primarily when the compressed medium shows a pressure deviating from the counterpressure in reaching the pressure side of the working opening.

The compressed medium flows alternately in and out of the chamber. The longitudinal waves arising in this connection can only be absorbed by complex and fluidic downstream components involving heavy losses. Subsequent aggregates are likewise caused to oscillate due to these pressure pulsations.

The constancy of the vibration frequency sometimes stimulates peripheral plant components to resonant vibrations. Therewith associated sound emissions sometimes then exceed permissible limit values.

Displacement compressors with chambers separated by blades provide the possibility of an irregular blade pitch, whereby the frequency of the pulsations fluctuates and resonances are avoided as far as possible in components near to the process.

In contrast to other displacement compressors, LRVPs possess a fluid piston. Pressure fluctuations in the gas phase to be compressed continue via the liquid ring and change its geometry and therewith also its centre of gravity.

As the centre of gravity periodically changes due to the rotation with constant speed, the resulting forces stimulate all component parts connected with the aggregate into vibration.

3.3.3
Vibration Stimulation by Flow Separations

With all types of compressors, limits are set to the compression realisable in one stage. The pressure ratios achievable with employment of LRVPs operating with water require multiple stages with employment of other oil-free compressors.

With LRVPs, the energy necessary for the compression of the gas and the pressure increase in the liquid are brought into the liquid ring via the blades.

The pressure difference prevailing between suction and pressure sides continues via the individual chambers. Since the blade areas are separated from each other only by a liquid, the dominant pressure gradient produces a flow via the blade tips. The cross-sectional constriction by the blade tips leads to an increase in the flow velocity and therewith also to a pressure drop. The pressure prevailing thereby at the blade tips can fall below the chamber pressure.

With increasing pressure difference between the working openings and also with declining rotation speed, the pressure is reduced at the blade tips. If this falls below the vapour pressure prevailing there, then the ring liquid evaporates. If the vapour bubble reaches the zone of higher pressure again, then it condenses abruptly. This procedure stimulates the blades to intense vibrations, which continue as impact sound via the other machine elements. Depending on operating condition, the flow separation leads to a significant sound emission.

3.3.4
Measures for Vibration Damping

3.3.4.1 Factory Measures
As much as possible, one takes factory measures for reduction of vibrations and sound emissions.

A well-engineered hydraulics and flow conveyance combined with sound-insulating materials for the employed components reduce and absorb the sound emission. Furthermore, rotating components of the LRVPs are balanced at Speck Pumpen dynamically on two levels according to DIN ISO 1940-1 with a balance quality G40.

3.3.4.2 Installation
Compared to other types of compressors, LRVPs are characterized by a smooth running. Nevertheless, if the operating performance with an installation corresponding to the operating instructions cannot satisfy the requirements, then peripheral plant components should be connected with the LRVPs mounted on the vibration dampers via suitable vibration compensators. Downstream surge tanks can once again reduce the amplitude of the pulsation by the compression.

3.3.4.3 Operating Mode

LRVPs should be operated with suction pressures below the double vapour pressure of the ring liquid delivered on the pressure side with a gas ejector. With the aid of the ejector, the gas to be sucked off is compressed and the LRVP is fed with higher pressure. The result is an increase in the suction capacity of the working point. Furthermore, the noise developments are prevented by flow separations.

The supplied quantity of operating liquid should not exceed the preset data. An overstepping of the presetting produces a back pressure of the liquid in the vacuum pump. This leads to an increase in the internal compression ratios and therewith also to the increase in the power consumption. The results can be an overloading of the components as well as an intensified noise development by flow separations.

With the extraction of the vapours condensable in the vacuum pump, attention is to be paid to a sufficient inert gas portion. Otherwise, intensified noise developments are to be expected by the condensation in the course of the compression likewise by flow separations.

3.4 Selection of Suitable Liquid Ring Vacuum Pumps

A goal of each new development and improvement is the creation of a marketable product, which, considering the operating conditions, produces as low total costs as possible. They consist of

investment costs (acquisition, start-up)
operational costs (consumption of operating resources, necessary service)

$$K_{Total} = K_{Invest} + K_{Operation}$$

Since the operational costs only accrue with the use of the products, these result from the specific costs and the service life:

$$K_{Operation} = k_{Operation} \cdot t_{Operation}$$

With increasing service life, the operational and therewith also the total costs rise. Taking into consideration the technical progress and the limited product life, it requires a consideration of the total costs over a limited time period, which at present is maximum 3 years. Thus, the service life constitutes the total time, which the product is operated in the 3 years.

This cost calculation makes it clear that with the product development, secure values over the service life and the specific operational costs must exist. For the product to be developed or to be improved always represents a trade-off between low operational and investment costs in favour of the total costs.

The constantly narrowing market and the rising energy costs make a more exact examination necessary in the choice and process connection of vacuum pumps. Only an optimal process connection makes possible a careful employment of the available resources. In view of the numerous process-specific details of LRVPs, the extensive cost analysis gains more and more importance.

An optimal employment always requires case-specific analysis. However, LRVPs optimised for concrete procedural processes should be introduced once in the following and their optimal process connection should be elucidated.

3.4.1
Simple, Robust and Suitable for the Entire Pressure Range

LRVPs in closed coupled design with variable pressure ports for covering the entire pressure ranging from approximately 40 to 1013 mbar represent at present the most frequently used design with new installations due to their good price–performance ratio. But even these vacuum pumps do not meet all the requirements. Thus, deposits in the dead spaces and drillings as well as defects of the valves of the variable pressure ports rank among the most frequent causes of failure.

Investigations of vacuum pumps with fluidic optimised working openings and free of dead space have shown that the named pressure range can also be covered without variable pressure slots.

Figure 3.13 shows the different behaviours of a vacuum pump, which was examined, on the one hand, with a conventional pump casing with inter casing and valve flaps and, on the other hand, with a fluidic optimised control cover. Equivalent suction capacities are to be established over wide ranges. However, the power requirement has decreased noticeably by the employment of the casing cover.

Furthermore, the comparatively large cross section and the missing dead space make the formation of deposits more difficult. Since it is missing, the suction capacity can no longer be affected by defective valves (Figures 3.14 and 3.15).

3.4.2
The Vacuum Pump for the Delivery of Liquid

The trend to higher efficiency of the energy employment requires a better sealing of the facilities and tools or even a reorganisation of the procedure in the case of vacuum processes. Consequently, the necessary throughput of vacuum pumps decreases. However, the liquids carried along or to be conveyed in many applications cannot for the most part be technically conditionally process reduced. The limited capacity of liquid conveyance of the vacuum pumps used primarily so far with level inter casing requires thus a separation of gas phase and liquid phase prior to entering the vacuum pump. The expenditure for technical instruments increases by at least one separator and a discharge pump.

Regarding the expenditure for technical instruments as well as the investment costs, the principle of the port cylinder represents an attractive alternative for gas–liquid separation.

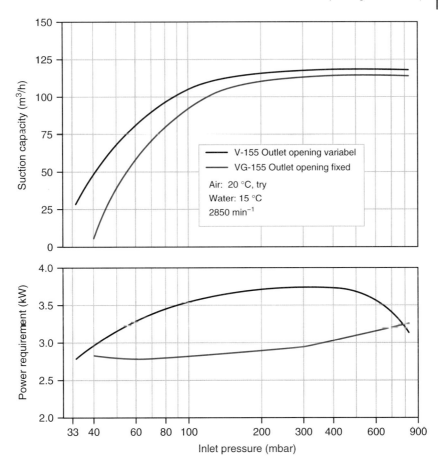

Figure 3.13 Characteristics of vacuum pumps with fixed and variable pressure slots.

With all LRVPs, the suction, compression and ejection occur on the basis of the movement of the liquid ring. The latter emerges and immerses again like a piston in the course of rotation of the blades. The liquid flows along with the movement in the circumferential direction radially to the hubs and back again.

With employment of an inter casing, the gas flows axially into the blade chamber or into the inter casing (Figures 3.2 and 3.3). For this purpose, an additional axial movement of the liquid is also necessary

With the port cylinder (Figure 3.16), the gas exchange takes place via the blade base through a pipe provided with working openings and a partition wall. With the hub control the working openings can extend over the entire length of the impeller, so that in comparison to the inter casing, significantly larger ports are available for the gas exchange.

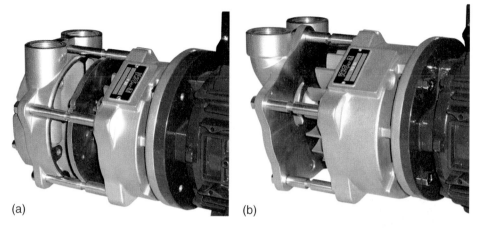

Figure 3.14 Design comparison with (a, V series) and without (b, VG series) dead spaces and valve technology.

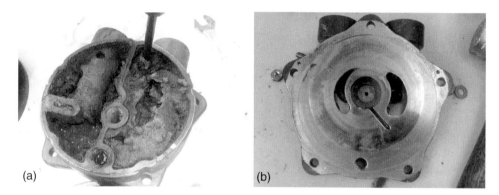

Figure 3.15 Solid deposits in the dead spaces of the V series (a) and (b) free channels in the VG series.

If a substantial liquid portion is to be conveyed, then the latter arrives in the case of port cylinders with comparatively less pressure loss via the working openings. With the employment of inter casings, the pressure loss is disproportionately higher. Already within the suction range, it rises due to the pressure loss of the liquid level in such a manner that the suction slot of the inter casing is partially occluded. Furthermore, the small pressure ports of the inter casing impede a free discharge of the liquid, so that the pressure between the blades rises far above the counterpressure.

Figure 3.17 shows impressively that, compared with other designs, the hub control enables even more liquid to be delivered. Even the larger pressure openings in the dead-space free design enable larger quantities of liquid to be handled.

3.4 Selection of Suitable Liquid Ring Vacuum Pumps

Figure 3.16 VN series with hub control.

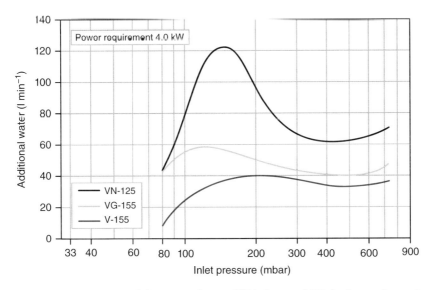

Figure 3.17 Comparison of the water tolerance (VN hub control; VG dead space free and V conventional).

It is to be inferred from the performance data described in Figure 3.18 that in the case of LRVPs with port cylinder, the suction capacity and the power consumption are affected only marginally by the water quantity conveyed. The port cylinder permits the conveyance of three to four times the liquid quantity compared to the

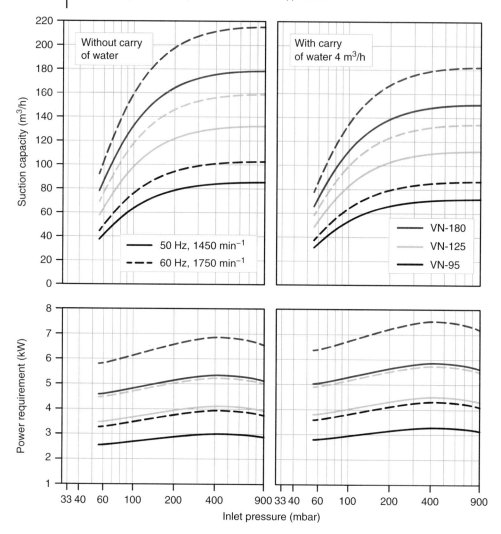

Figure 3.18 Influence of additional water on the performance data of a liquid ring vacuum pump with port cylinder.

inter casing, so that with the preponderant number of applications a gas–liquid separation prior to entering the vacuum pump can be dispensed with.

Since with the employment of this vacuum pump the volume portion of the liquid is comparatively high on the process side in the medium to be pumped, the suction and pressure lines must be arranged constantly downwards in the direction of the flow. Furthermore, on the pressure side, the medium must run out freely into a separator. Otherwise liquid collects in the sinks, so that the gaseous phase can only be fed discontinuously (Figure 3.19). This leads to a fluctuating vacuum depending on the supply of liquid.

3.4 Selection of Suitable Liquid Ring Vacuum Pumps

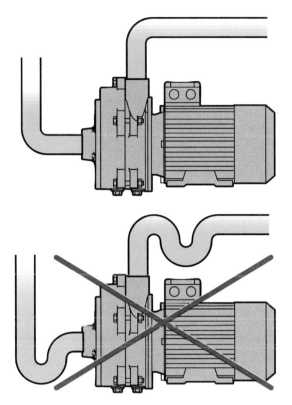

Figure 3.19 The correct installation situation for gas–liquid mixtures with high proportion of liquid.

3.4.3
Quiet and Compact for a Vacuum Close to the Vapour Pressure

Increased noise emission is an indicator of a low degree of efficiency. Also, the generation of vibrations as the source of the sound emission requires an application of power. Furthermore, the components are unnecessarily stressed by the vibrations. Thus, the sound emission therefore has a direct impact on the operating costs, on the one hand, through an increased power requirement and, on the other hand, through a high service expense.

A cause of the noise emissions most frequently to be encountered are separation noises via the operation close to the vapour pressure of the ring liquid or with a pressure difference which is too great. Previous experiences in numerous applications have shown that over wide ranges of two-stage LRVPs tend less to separation noises. However, they were replaced in numerous applications for lack of space and for reasons of economy by single-stage closed coupled pumps.

Often, one can no longer satisfy the changes of the process controls as well as the stricter requirements for emissions with the previously used closed coupled vacuum pumps. Frequently, the space and the necessary financial leeway are lacking

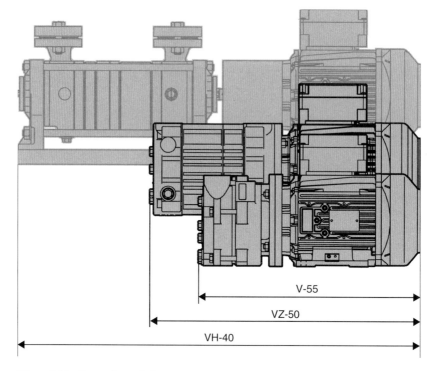

Figure 3.20 Comparison of sizes.

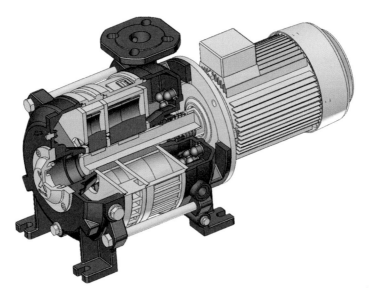

Figure 3.21 Section of a two-stage vacuum pump in closed coupled design.

for the reintroduction of the traditional two-stage LRVPs in the optimised devices and facilities.

Best suited for these applications are optimised two-stage LRVPs in closed coupled design. They combine the advantages of the traditional two-stage vacuum pumps with those of the closed coupled machines and set new standards with regard to noise emission as well as in the price–performance ratio (Figures 3.20–3.22).

3.4.4
A Compact Unit

Usually, LRVPs themselves cannot be used directly for creating vacuum. They must be used together with other components to create a unit. Units of this kind are now available as serial products in larger quantities, but are still based on the elaborate 'plant engineer concept'.

There is also a much more compact and low-cost option – based on the 'pump manufacturer concept'. This concept has been tried and tested for decades in oil-lubricated rotary piston vacuum pumps – so why not in LRVPs. There is no need for extensive frame constructions for mounting system components. Cost-intensive pipes required for manufacturing previous units, as well as throughout the service life, are no longer required. Components can be retrofitted easily.

Figures 3.23 and 3.24 show compact units, which contain both an LRVP with a separator and a heat exchanger. Liquid level monitoring and cavitation protection are features that can be implemented extremely easily.

3.4.5
The Right Sealing Concept

LRVPs are still frequently used to be called water ring pumps, even though they have been used in the chemical industry since the start of the last century. In this sector, they are not only used for creating vacuum but also used as reactors, absorbers, condensers and washers. Process liquids are being used ever more frequently to achieve higher product qualities and pure products.

However, it is not always easy to operate LRVPs with some of the process media used. Depending on the volatility, ignitability and toxicity of the ring liquid, as well as the delivered gases and vapours, measures must be implemented, which, on the one hand, guarantee the high quality of the product and, on the other hand, pose no risk to colleagues and the environment. The leak-tightness of the vacuum pump is a key aspect here. The so-called dynamic seals, which have to seal a rotating shaft against the stationary housing, for example are critical in this respect.

3.4.5.1 Single-Acting Mechanical Seal
Single-acting mechanical seals are still most frequently used for this task. The functional principle means that there is no 100% leak-tight mechanical seal, as the rotating sliding surface is lubricated and carried by the process medium. This

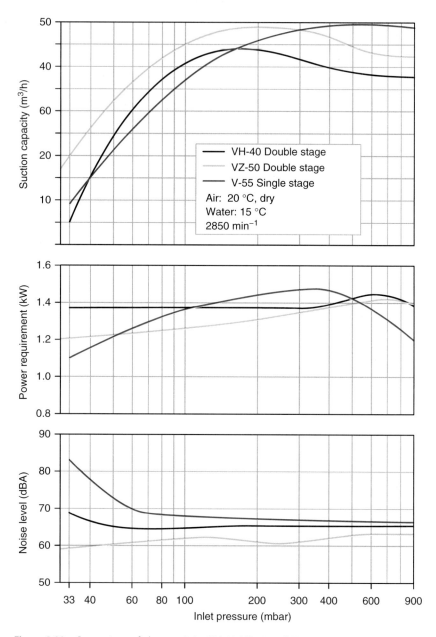

Figure 3.22 Comparison of characteristics VH-40, VZ-50 and V-55.

Figure 3.23 Single-stage vacuum pump with heat exchanger.

Figure 3.24 Two-stage vacuum pump with heat exchanger and precondenser.

results in small quantities of the process medium passing through the sealing gap (Figure 3.25).

3.4.5.2 Double-Acting Mechanical Seal

One option for preventing process medium from escaping is to use double-acting mechanical seals. As the name suggests, two mechanical seals are mounted one behind the other. With this mechanical seal, a non-hazardous liquid is pushed between the two seals to lubricate the sliding surfaces. A higher pressure in the lubricating fluid compared with the process media prevents leaks through the seal to the surrounding area. However, this also means that low quantities of lubricating fluid are pressed into the process. It is therefore important to choose suitable medium combinations during the system planning process. During operation, relevant measurement and control technology must always be in place to provide

Figure 3.25 Section through a vacuum pump with single-acting mechanical seal.

Figure 3.26 Section through a vacuum pump with double-acting mechanical seal.

adequate pressure overlap and cooling (friction heat on the sliding surfaces) of the lubricating fluid (Figure 3.26).

3.4.5.3 Magnetic Coupling

There are no problems with leaks with a magnetic coupling. No dynamic seal is required for this design. The forces are transferred through the wall of a containment shell via magnetic fields. But there is also a negative side to this system. The rotating components in the vacuum pump must now be positioned in the process medium. Slide bearings are used for this, which are lubricated and cooled with the

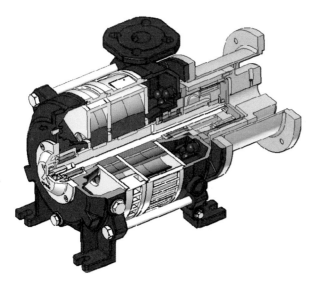

Figure 3.27 Section through a vacuum pump with magnetic coupling.

ring fluid. Solid components and liquids with low viscosity or which evaporate easily under process conditions reduce the service life of the slide bearings and endanger the process stability. An otherwise normal and unproblematic rotation test on the dry vacuum pump can destroy the bearings (Figure 3.27).

3.4.6 Vacuum Control

The goal of the vacuum control is to maintain a specified vacuum, which can be constant as well as variable with regard to time.

3.4.7 Valve Control

With valve control, the actual and target vacuum is adjusted by influencing the suction rate through valve technology. For this, one or multiple vacuum pumps are operated constantly at a fixed speed.

3.4.7.1 Bypass Recovery

Bypass recovery is still the most widely used form of vacuum control today for LRVPs. In this process, an already pumped gas is released from the pressure side to the suction side via a valve. If the process vacuum is lower than desired, the valve is opened until the setpoint value is reached.

As, for the most part, gases are sucked off, which are released directly into the ambient air following the suction process, ambient air is also used for pressure control (Figure 3.28).

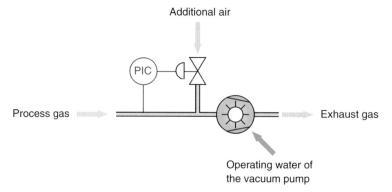

Figure 3.28 Vacuum control through false air supply.

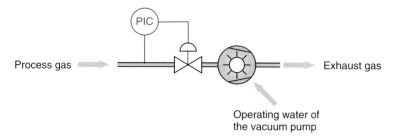

Figure 3.29 Vacuum control by reducing the suction flow.

3.4.7.2 Reduction on the Suction Side

Direct reduction of the pumped process gas prior to entering the vacuum pump is possible but seldom used in practice in relation with LRVPs. The gas is relaxed following the reduction. This results in a larger pressure difference on the vacuum pump, which reduces the suction capacity (Figure 3.29).

It is important to remember, however, that reduction is only possible to a limited extent here. Excessive reduction results in the liquid ring vacuum pump being operated in the cavitation range.

3.4.8
Power Adjustment

The valve control enables the vacuum to be set precisely, but is highly inefficient with regard to energy in most cases. Gas is reduced unnecessarily, only to compress it again later without making use of the energy consumed.

One solution that requires a higher investment, but is more efficient and low cost to run, is to adjust the suction power of the vacuum pumps.

3.4.8.1 Cascade Connection of Vacuum Pumps

One option for adapting the vacuum power to the requirements of the process is to connect and disconnect a set of vacuum pumps connected in parallel. Connecting and disconnecting the vacuum pumps causes slight fluctuations in the vacuum depending on the system volume. This must be taken into account during planning.

This type of switching method is useful if the volumes to be sucked off fluctuate considerably when the vacuum pumps are in operation. Only then can vacuum pumps be switched off to reduce the energy consumption by a fraction (Figure 3.30).

3.4.8.2 Speed Adjustment in Vacuum Pumps

Another option for adjusting the power is to change the speed of the vacuum pump (Figure 3.31).

Increasing the speed over the rated speed can be used to increase the suction capacity almost proportionately to the speed. The power requirements increase disproportionately during this process.

The limit usually lies at approximately 70 Hz for units designed for 50 Hz AC. The strain caused by the higher torque and circumferential speeds is still unproblematic here.

Reducing the suction power by lowering the speed results in a disproportionate drop in the power requirements. However, it must also be taken into account, as shown in Figure 3.32, that the achievable vacuum falls over vast areas. The possible speed reduction in this case, therefore, depends on not only the required suction power but also the vacuum, which can be achieved with the vacuum pump.

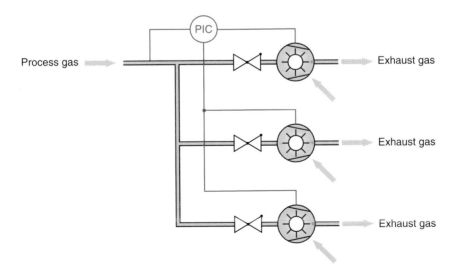

Figure 3.30 Vacuum control via cascade connection of vacuum pumps.

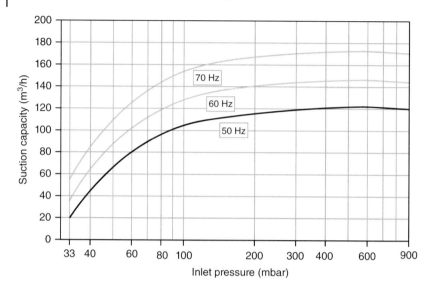

Figure 3.31 Vacuum control by changing the speed; speed > rated speed.

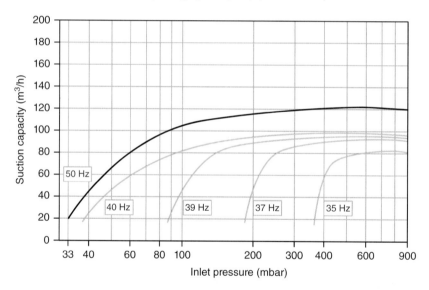

Figure 3.32 Vacuum control by changing the speed; speed < rated speed.

The speed adjustment also enables the suction power to be adjusted extremely precisely and should be used if the required suction power lies in the control range of the vacuum pump.

3.4.8.3 Cascade Connection Combined with Speed Adjustment

In many cases, multiple speed-controlled vacuum pumps are also used together. This is necessary in situations where cascade connection with a precision as achieved with speed adjustment is required and the suction capacity requirements fluctuate considerably.

3.5
Process Connection and Plant Construction

3.5.1
Set-Up and Operation of Liquid Ring Vacuum Pumps

With the set-up and operation of LRVPs, the following aspects should be considered without fail in the interest of as long a running time as possible:

- After connection of the engine, the *direction of rotation* is to be checked. If the vacuum pump is filled to half of the suction chamber volume (never more), then with wrong direction of rotation the operating liquid is discharged via the intake port. With a still dry LRVP, one should start the engine only briefly to check the direction of rotation (approximately 1 s), since otherwise the slip ring sealings are damaged.
- If the operating liquid is fed via a *pressure line*, then pressure adjacent to the pump should not fall below *0.2 bar positive pressure*. Furthermore, for the prevention of an overfilling during stoppage, the *liquid supply* coupled with the motor is *to be connected or disconnected*.
- If the vacuum pump is fed with operating liquid from an adjoining tank, then the *liquid level in the tank* should always correspond to about the *shaft height* of the compressor. Furthermore, it is to be taken into consideration that if possible the *pressure loss* in the supply line for the vacuum pump with a maximum vacuum should not exceed *100 mbar*.
- If for lack of space and for reasons of economy only a set-up separator is used, then one can prevent an overfilling above the wave height by installation of an automatic *drain valve*.
- If a *substructure separator* is employed, then the *vacuum pump* is also to be *replenished* before the initial operation with liquid up to the shaft height. An emptying into the tank can be prevented by a suitable *check valve* in the operating water supply to the compressor.
- Since a compression effect can first take place after assembly of the liquid ring, the vacuum pump may *not be started up towards the closed suction side*. And also an appropriately adjusted *vacuum check valve* offers sufficient security
- On behalf of stable pressure ratios, the pressure-sided line leading to a separator should always be run horizontally or *downwards* in the direction of the flow.
- LRVPs are often used in facilities, in which liquid is conveyed by the gas to be suctioned off. For the prevention of pressure fluctuations, these lines are also to be run horizontally or *downwards* in the direction of the flow.

3.5.2
Conveyance of the Operating Liquid

Due to the intensive phase contact between the ring liquid and the medium to be absorbed, the compression takes place almost isothermally, that is the temperature gas changes only marginally. Thus, almost the entire shaft power supplied as well as the heat brought in via the incoming media must be discharged in the form of heat via the operating liquid.

For the realisation of as high a degree of efficiency as possible, the employment of as large and cold an operational resource flow as possible was therefore necessary. One attains a maximum operating liquid flow, if the heated liquid is completely discharged. Here one speaks of throughput cooling (Figure 3.33).

Sometimes, water is not available for an unlimited period and represents a cost factor to be taken into account. It is to be inferred from Figure 3.9 that the influence of the operating water temperature decreases with rising suction pressure, that is the water with low intake pressure should be heated up as little as possible. Therefore, a larger water flow is necessary.

With high intake pressure, on the other hand, a stronger heating leads only to a very small loss, so that the water quantity could be reduced taking into account

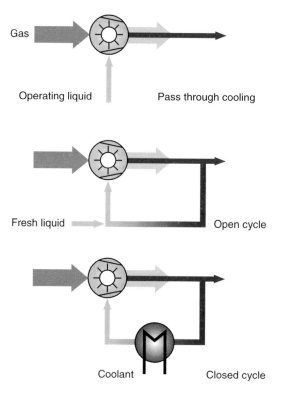

Figure 3.33 Operating modes of liquid ring vacuum pumps.

the total costs. However, since a set liquid supply is necessary for the maintenance of the hydraulic ratios in the LRVP, a reduction of the water requirement can only occur by the refeeding of a part of the water already ejected. In this case, one speaks of an open circuit cooling (Figure 3.33).

If no open cooling circuit is possible due to the structural conditions of the procedural processes, then the ring liquid is completely returned. The heat is then to be dissipated via an additional heat exchanger. It is now a matter of a closed circuit cooling (Figure 3.33).

3.5.3
Precompression

Due to the principle of operation, the working range of LRVPs is limited by the vapour pressure of the ring liquid. However, if a vacuum is necessary below the double vapour pressure of the operating liquid, then a further compressor must be connected upstream, with the aid of which one can realise lower intake pressures.

However, the precompressor used should not detract from the advantages obtained by the employment of the LRVP. Among these are, for example a medium-neutral compression by prevention of additional lubricants, the insensitivity towards droplets and solid particles that are carried along, and much more.

The most frequently used are gas ejectors, steam jets with downstream condenser and rotary piston compressor.

3.5.3.1 Precompression by Gas Ejector
The use of gas ejectors for precompression is the simplest variant for the generation of a vacuum below the double vapour pressure of the ring liquid (Figure 3.34).

Principle of Operation The power necessary for the compression is first made available by the ambient air. The relaxation of the ambient air to the suction pressure in a Laval nozzle makes possible the acceleration of the gas particles to ultrasonic velocity. The comparatively slow particles of the extracted medium experience an

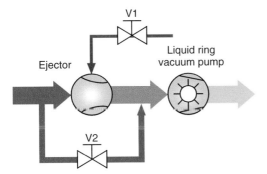

Figure 3.34 Arrangement of a combination of gas ejector and liquid ring vacuum pump.

increase in velocity through the molecular interaction with the gas propellant in the mixing tube. A compression by conversion of the kinetic energy of the gas mixture into pressure energy follows the mixing process. The compressed suction medium together with the propellant air is fed to the LRVP. In the vacuum pump, in addition to the compression of the suction medium, a compression also now takes place of the previously relaxed propellant air to the ambient pressure (Figure 3.35).

If one supplies the propellant gas from the already compressed inert gas flow, then the system is in itself closed, and for this reason, it can be used for applications without air supply.

On the basis of the precompression, the liquid ring compressor can now be operated with higher intake pressure. With a suction pressure at the gas ejector close

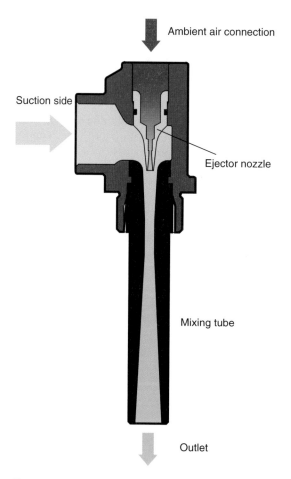

Figure 3.35 Design of a gas ejector.

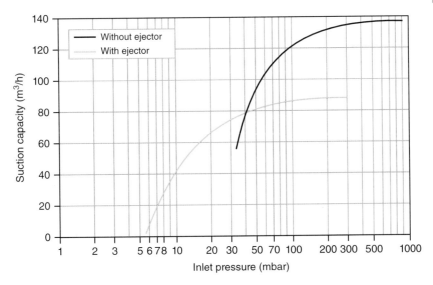

Figure 3.36 Pump characteristics with and without gas ejector.

to the boiling pressure of the ring liquid, a substantially higher intake volume can then be obtained (Figure 3.36).

With further divergence of the boiling pressure of the ring liquid, the gas entering via the propellant nozzle as well as the restrictor losses in the narrow mixing tube produces a decrease in the suction capacity of the combination, so that an operation without gas ejector makes sense.

In each case, the maximum suction capacities can be obtained by the installation of a bypass parallel to the gas ejector according to Figure 3.34. If, for example with the evacuation of a tank up to a pressure of 42 mbar (cf. Figure 3.36) the valve V1 remains closed and V2 opened (cf. Figure 3.34), then the greater suction capacity of the vacuum pump is first available. If valve V2 is closed now and valve V1 opens, in the further course of the evacuation the greater suction capacity of the combination of gas ejector and vacuum pump can be utilised.

The exact intersection point of the characteristics and thus the switching point for the valves depend on the respective operational conditions.

The pressures and suction capacity obtainable with the aid of gas ejectors are limited by the suction capacity of the vacuum pump, since the propellant gas used is always to be compressed once again.

3.5.3.2 Precompression by Steam Jet

If greater suction capacities with low pressures are to be realised, then with the availability of water vapour, steam jets are often used. The vapour used as propellant gas can still be thermally compressed in a condenser prior to entering the

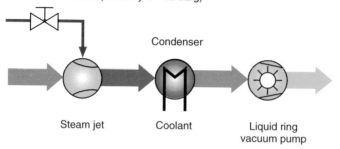

Figure 3.37 Arrangement of a combination of steam jet and liquid ring vacuum pump.

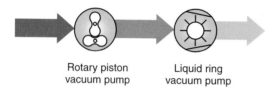

Figure 3.38 Arrangement of a combination of rotary piston and liquid ring vacuum pump.

LRVP. Limits are set here by the available cooling water. As a rule, one can discharge the incidental condensate with the LRVP (Figure 3.37).

3.5.3.3 Precompression by Rotary Piston Vacuum Pumps

From the point of view of economy, rotary piston compressors are sometimes also used for precompression. These volumetrically working, two-phase, dry-running rotary piston compressors are characterized by insensitivity towards impurities in the medium to be pumped.

The energy necessary for the compression is provided by a motor drive. Due to the significantly greater degree of efficiency of rotary piston compressors, a smaller total drive power suffices in contrast to gas ejectors in combination with liquid ring compressors (Figure 3.38).

It is to be taken into account that with the employment of rotary piston vacuum pumps, the temperature of the fluid fed to the liquid ring compressor is higher than with the other combinations. In addition, this heat is to be dissipated by the ring liquid, so that the temperature of the liquid ring will be higher and with it the suction capacity will be lower.

Which of the addressed combinations can be used in detail, must be clarified in detail first technically and then by an amortisation calculation. Depending on operating conditions, the one or the other variant can be the most economical.

3.6
Main Damage Symptoms

3.6.1
Water Impact

The most frequent cause of failure by incorrect installation is the too high or even total flooding of the LRVP. This is caused by

- missing cut-off of pressure water;
- neglect of condensate entry during stoppage;
- neglect of liquid accumulated in the intake lines;
- liquid level in the side-mounted separator is too high;
- start-up under vacuum;
- non-permissible counterpressure.

Above all, with two-stage designs, the flooding leads to the destruction of the impeller of the second stage. The water inserted by the first stage cannot be extracted in the second stage because of the smaller chamber volume. In most cases, a torn blade destroys the inter casings of the second stage as well as bearing and floating ring seal on the pressure side (Figure 3.39).

3.6.2
Cavitation

3.6.2.1 Concept
Different definitions can be found in the literature for the concept of cavitation, in which the following statements, however, combine in essence (Figure 3.40):

Figure 3.39 Impeller of the second stage after a flooding.

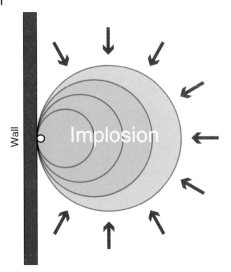

Figure 3.40 Cavitation in a germ.

> *If with a vaporous medium one exceeds the boiling curve by a lowering of the temperature or increase of the pressure, then the vapour condenses to so-called germs. Droplets, particles, and solid surfaces serve as germs. If sufficient germs are not available, then the vapour is supercooled. If it then obtains a germ, then it condenses abruptly. In consequence of the developing high pressure differences in the chamber, the then emerging high flow velocities lead to a damaging of the solid surfaces present in the nearer environment.*

Similar to the descriptions in Figure 3.41, the ring liquid is accelerated by the cavitation into the space which is becoming open due to the condensation. Thus, the liquid ring reaches the hub before the conclusion of the volume decrease and almost fills the whole area between the blades.

In the course of the further decrease in the available space by rotation of the impeller, the water can only escape along the direction of rotation of the impeller. Depending on the intensity of the cavitation, the backflow resulting therefrom leads to an audible flow separation at the blade tips.

Typical cavitation damages are like cavities in the solid caused by pinholes (Figure 3.42).

The destructive effects of cavitation can be prevented as far as possible if a sufficiently high portion of inert gas is present. The non-condensable components, despite the still abrupt condensation of the vapour, prevent a development of larger pressure gradients and thus the production of high velocities responsible for the destruction of the two-phase flow.

As a rule of thumb, the suction pressure should not fall below twice the vapour pressure equilibrium of the liquid. At the water temperature of 15 °C, the corresponding vapour pressure equilibrium of 17 mbar corresponds to a minimum section pressure of approximately 34 mbar.

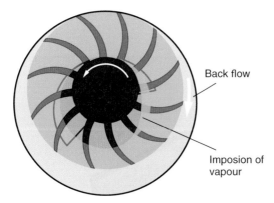

Figure 3.41 Cavitation in liquid ring vacuum pumps.

Figure 3.42 Cavitation damages to an inter casing.

With the employment of LRVPs, sometimes the destructive cavitation produces a strong noise development by flow separation. However, this is unsuitable as an indication for the prevention of damages by cavitation, since it does not occur compellingly and beyond that can also have other causes.

3.6.3
Calcareous Deposits and How to Avoid Them

3.6.3.1 Water as Service Liquid in Liquid Ring Vacuum Pumps (LRVPs)
Due to its high availability, environmental impact, non-toxicity and excellent thermophysical material properties, water is one of the most important and most widely used compounds and therefore also the most frequently used service liquid in LRVPs.

Figure 3.43 Calcinations.

However, water is not found chemically pure in nature, but rather contains a multiplicity of dissolved materials, such as salts, acids and gases, which significantly shape the formation of deposits as well as the corrosion behaviour of water in technical use.

On behalf of a satisfactory service life of the facilities and aggregates operated with water, a monitoring of the water quality makes sense above all with open systems (new fresh water is injected and after one cycle discharged).

The deposit and corrosion behaviour of water (excepting chemically pure water) is primarily determined by the carbonic acid content in the water. Thus, a carbonic acid content that is too high leads to a stock removal of the so-called black ferrous materials. A carbonic acid content that is too low results in deposits by precipitation of lime (Figure 3.43).

Calcifications in LRVPs lead to a rise in the power consumption and sometimes interfere substantially with the suction capacity by constriction of the throughput cross section.

3.6.3.2 Prevention by Monitoring the Water Quality

Since the deposit and corrosion behaviour of water are determined primarily by the carbonic acid content in the water, the evaluation of carbonate hardness and of the pH value permits a practical assessment of the significance of the water quality for the service life of the LRVP.

Measurement Categories for the Determination of Water Quality The carbonate hardness is a measure for the alkaline earth ions bound as bicarbonates [$Ca(HCO_3)_2$ and $Mg(HCO_3)_2$] dissolved per litre water. These alkaline earth bicarbonates are mainly responsible to be a reason for unwanted deposits.

The higher the concentration of alkaline earth ions in water is, the higher is the 'hardness' of water. Water can be classified into different degrees of hardness, depending on the concentration of alkaline earth ions in it, measured in $mmol\, l^{-1}$:

<1.3 mmol l^{-1}	Soft	>2.5–3.8 mmol l^{-1}	Hard
>1.3–2.5 mmol l^{-1}	Medium hard	>3.8 mmol l^{-1}	Very hard

The pH value is a measure of the concentration of the hydrogen ions in the water. Thus, for example hydrogen chloride dissolved in water breaks down into its ions, whereby the portion of hydrogen ions increases.

$$HCl + H_2O \leftrightarrow (H_3O)^+ + Cl^-$$

Injected caustic soda likewise breaks down into its ions. However, here a part of the hydroxyl ions combine with the hydrogen ions. The portion of hydrogen ions decreases.

$$NaOH \leftrightarrow Na^+ + OH^- \quad (H_3O)^+ + OH^- \leftrightarrow 2\,H_2O$$

With the help of the pH value, the water quality can be subdivided regarding the reaction with other components entering via the process.

The definition of pH value is as follows:

$$pH = -\log c\,(H_3O)^+$$

pH value	Reaction	Portion of $(H_3O)^+$ ions
<3	Strongly acidic	Very high
4–6	Weakly acidic	High
7	Neutral	—
8–10	Weak alkaline	Small
11–14	Strong alkaline	Very small

Deposit and Corrosion Behaviour As consequence of the preceding details, there are trends in opposite directions with regard to the deposit and corrosion behaviour.

On the one hand, sufficient carbonic acid must be available for the solution of the carbonates, so that the process is not impaired by deposits. On the other hand, on behalf of the prevention of a corrosive degradation, the carbonic acid portion should not be too high, so that a passive coating can be formed on the ferrous walls of the LRVP by forming slight deposits.

The carbonic acid concentration with which neither the formation of a passive coating nor the precipitation of lime occurs is plotted in the following diagram as an equilibrium curve (Figure 3.44).

Since no precise calibration of the pH value is possible on the equilibrium curve with most applications which are determined process and plant specifically, taking into consideration the flow of the material supplied in the temporal medium a pH value should be aimed at if possible slightly above the equilibrium curve.

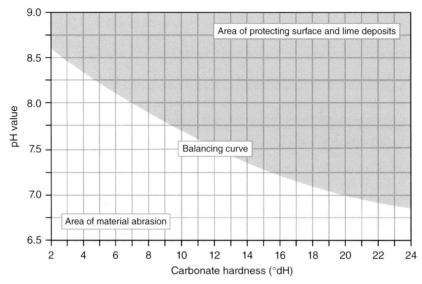

Figure 3.44 Deposit and corrosion behaviour of water at 17 °C, in particular valid for black ferrous material.

3.7
Table of Symbols

Symbols			
	a	$\mathrm{m\,s^{-2}}$	Acceleration
	c	$\mathrm{mol\,l^{-1}}$	Concentration
	C	$\mathrm{kJ\,(kg\,K)^{-1}}$	Specific heat capacity
	F	N	Power
	H	$\mathrm{kJ\,kg^{-1}}$	Specific enthalpy
	K	$\mathrm{J\,K^{-1}}$	Boltzmann constant
	m	kg	Mass
	M	$\mathrm{kmol\,kg^{-1}}$	Molar mass
	n	kmol	Molar quantity
	p	Pa	Pressure
	P	kW	Power output
	Q	kJ	Heat quantity
	r	m	Radius
	R	$\mathrm{kJ\,(kmol\,K)^{-1}}$	Universal gas constant
	t	°C	Temperature
	T	K	Temperature
	V	$\mathrm{m^3}$	Volume
	ω	$\mathrm{rad\,s^{-1}}$	Angular velocity

Indices	1, 2, … *n*	Counter
	D	Vapour
	F	Ring liquid
	G	Inert gas
	Hydr	Hydraulic
	K	Catalogue value
	P	Isobar
	R	Rotation/rotor
	S	Condition at the inlet nozzle
	V	Compression
	T	Carrier
	W	Shaft
Other marking	→	Vector
	·	Time derivative

4
Steam Jet Vacuum Pumps

Harald Grave

4.1
Design and Function of a Jet Pump

Jet pumps are devices for the transfer, compression or mixing of gases, vapours, liquids or solids, in which a gaseous or liquid medium acts as the motive force. They operate by the conversion of pressure energy into velocity by means of suitable nozzles. They are 'pumps without moving parts' (Figure 4.1).

The basic principle of the jet pump is that the liquid or gas jet exits the motive nozzle at a high velocity and low pressure and entrains and accelerates the surrounding liquid, gas or solid medium. The result of this action is the mixing of the driving and the entrained material at a mean velocity. In a second nozzle, this velocity of the mixture is reduced and the pressure increases to an outlet pressure that is higher than the suction pressure.

Vacuum Technology in the Chemical Industry, First Edition. Edited by Wolfgang Jorisch.
© 2015 Wiley-VCH Verlag GmbH & Co. KGaA. Published 2015 by Wiley-VCH Verlag GmbH & Co. KGaA.

4 Steam Jet Vacuum Pumps

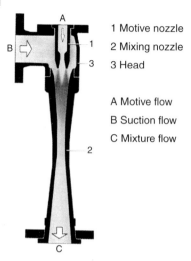

1 Motive nozzle
2 Mixing nozzle
3 Head

A Motive flow
B Suction flow
C Mixture flow

Figure 4.1 Steam jet compressor.

The conversion of this principle to a practical purpose requires a simple apparatus, which normally consists of only three main parts: the motive nozzle (1), the diffuser (2) and the head (3). There are three external connections to be designated: the motive medium connection (A), the suction nozzle (B) and the pressure nozzle (C).

Steam jet pumps are frequently used in the chemical industry for vacuum generation. The main reasons for this are their high operational reliability, their resistance to corrosion, their tolerance of fouling as well as their suitability for very large volume flows. However, when jet pumps are operated with steam as the motive medium, there will inevitably be waste water generated because the motive steam must be condensed.

In Figures 4.2 and 4.3, different types of jet pumps illustrate the wide range of applications, designs and materials of construction.

Figure 4.2 First stage of a steam jet vacuum pump for a steel degassing plant.

Figure 4.3 Overview of different types of jet pumps.

Figure 4.4 shows the thermodynamic processes in a jet pump in a *Mollier* enthalpy–entropy diagram. The inlet conditions of motive flow and suction flow are assumed to be saturated steam. This is also a common situation in real processes. Besides this, it is also assumed that the expansion in the motive nozzles ends just with the suction pressure p_0. In the case of a non-dissipative expansion in the motive nozzle, the change of state would be described by the perpendicular from point 1 to point 2. While a real expansion with losses leads to point 2'.

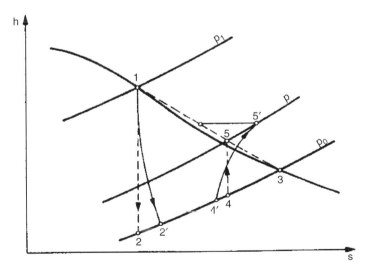

Figure 4.4 Changes of state in a steam jet compressor, shown in h,s-diagram (h = enthalpy and s = entropy).

It can be seen that the end point of the expansion (point 2′) is in the area of wet steam. Thus small droplets of condensate will form in the motive jet. In the motive nozzle, combined with the decrease of pressure, the temperature will also decrease. In the wet steam region the steam is cooled to the boiling point temperature corresponding to the pressure. When the motive steam expands to a pressure lower than 6 mbar, the corresponding temperature is below 0 °C, thus ice will form. Steam jet vacuum pumps for such applications are often heated at the mixing nozzle and sometimes also at the motive nozzle, to prevent the ice crystals from adhering to the internal wall. This would cause a constriction of the cross-sectional area and adversely effect the flow.

The motive nozzle is shaped like a *Laval nozzle*. This means there is an enlargement of the diameter after the smallest cross section. This is necessary to achieve velocities higher than sonic speed. For steam an expansion pressure ratio of only $p_1/p_0 = 1.73$ is sufficient to just achieve sonic velocity (critical pressure ratio). At higher expansion ratios (supercritical pressure ratios), the exact critical pressure and sonic speed is achieved in the smallest cross section. In these cases in the divergent part of the motive nozzle, a supersonic velocity results from a continuing expansion. Owing to the 'blocking' of the velocity to the sonic speed in the smallest cross section, the mass flow rate of such a supersonic nozzle only depends on the state of the motive media in front of the nozzle and of course on the diameter d_1. Here the mass flow rate m_1 is proportional to the motive pressure p_1.

$$m_1 = \text{konst.} \cdot d_1^2 \cdot p_1 \tag{4.1}$$

In the inlet cone of the mixing nozzle, the motive flow of state 2′ and the suction flow of state 3 are mixed to the mixture flow of state 4′. This happens at nearly constant pressure. In most cases, the velocity of the mixture flow is higher than sonic speed. When such a flow is decelerated, a shock wave will form with an abrupt increase of pressure and temperature. This shock wave is formed in the smallest cross section of the diffuser, when the ejector is working under design conditions. If the operating conditions differ from the design values, particularly when the discharge pressure p is lower than designed, the shock wave can move into the divergent diffuser. The temperature increase can often be felt by hand on the outside of the ejector. The change of state across the shock wave and in the following diffuser gives the point 5′ in the h,s-diagram.

The run of the pressure profile along the jet compressor is shown in Figure 4.5.

4.2
Operating Behaviour and Characteristic

In Figure 4.6, an example for the suction characteristic of a jet compressor is plotted. It shows the correlation between the suction pressure p_0 and the suction mass flow rate m_0.

Over a wide range, this characteristic shows that pressure and mass flow rate are directly proportional. An extrapolation of this straight line would reach the

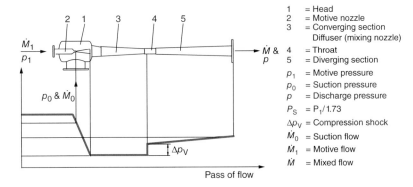

Figure 4.5 Schematic pressure profile in a jet compressor.

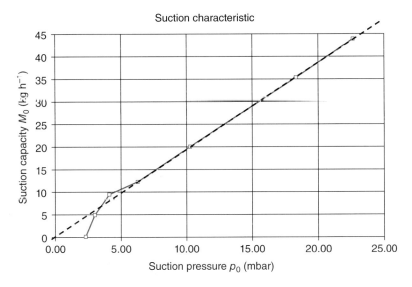

Figure 4.6 Suction characteristic of a gas- or steam jet compressor.

point of origin of the axes. This means that this part of the characteristic can be described by $m_0 = $ const. p_0. Because the ideal gas equation says $p_0 V_0 = m_0 R T_0$, the volume flow rate also has to be constant here. But for $m_0 = 0$ the suction pressure cannot be $p_0 = 0$. Therefore, the curve in Figure 4.6 shows a turning point for low suction pressures and for $m_0 = 0$ a definite end vacuum is achieved.

As well as the suction pressure, the discharge pressure p also has an influence on the operating behaviour. At this point the motive pressure is assumed to be constant. In Figure 4.7, the run of the curve for the suction mass flow rate is illustrated for a constant value of the suction pressure but for increasing discharge pressure.

For a wide range, no influence of the discharge pressure can be seen. This is caused by the supersonic flow in the mixing nozzle. Not until the increasing discharge pressure reaches the value of the critical discharge pressure, is the influence

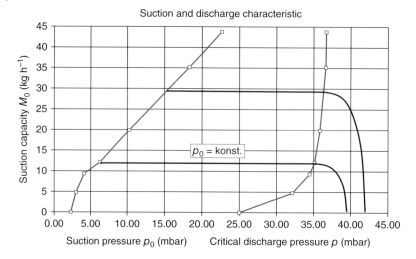

Figure 4.7 Influence of the discharge pressure on the suction capacity, for constant suction pressure.

of the pressure at the outlet of the ejector is noticeable. In Figure 4.7, this can be seen by the decrease of the suction mass flow rate from this point on. In real operation this means that an efficient and smooth function of the jet pump is only possible when the discharge pressure is lower than the critical value, stated in the specification. For a jet compressor with given dimensions, a family of curves $m_0 = f(p)$ with the suction pressure as a parameter can be drawn (see Figure 4.7).

In Figure 4.8, the influence of the motive pressure is shown. The increase of the motive pressure from 4 to 6 bar mainly causes a displacement of the critical

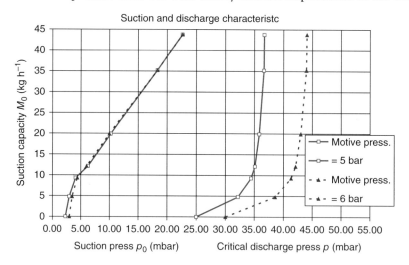

Figure 4.8 Influence of the motive pressure on the suction characteristic of a jet compressor.

discharge pressure curve. This shift has nearly the same ratio as the increase of the motive pressure. The suction pressure curve is only changed marginally.

The increase of the motive pressure does not only change the pressure ratio along the motive nozzle, but also the mass flow rate calculated by Eq. (4.1) rises. A similar effect to the increase of the motive pressure results from an increase of the motive mass flow rate, by increasing the diameter of the motive nozzle. In this case, the values for the critical discharge pressure also grow, while the suction characteristic is virtually unchanged.

4.3
Control of Jet Compressors

A simple method to influence the operational behaviour of a jet compressor is to throttle the motive pressure by a control valve. With a control as shown in Figure 4.9, the motive steam pressure can be reduced and thereby the motive mass flow rate is similarly reduced. This means that for lower motive flows the critical discharge pressure is also reduced. But as previously explained, this has no direct influence on either the suction capacity or the suction pressure, as long as the actual discharge pressure is lower than the critical one.

Therefore, the command variable is normally the actual discharge pressure. Such control systems are mainly installed with jet compressors or jet vacuum pumps that discharge into a condenser. When the cooling water temperature fluctuates, the condenser pressure also changes. In addition, this means the actual discharge pressure of the jet pump changes and the motive mass flow rate can be controlled depending on the condenser pressure. The advantage is to save steam when the cooling water temperature and thus the condenser pressure are lower than the design value. For example, if a design was made for hot summer

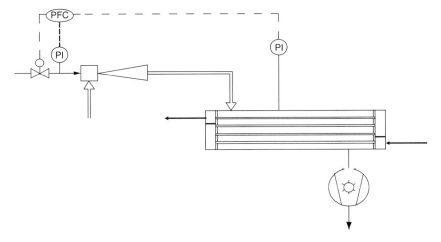

Figure 4.9 Control of a steam jet compressor (steam saving control).

temperatures, and without such a control, on cold winter days with lower cooling water temperatures the motive steam consumption would be much higher than necessary.

For special conditions it is also possible to control the motive steam pressure depending on the suction pressure. Such an installation is illustrated in Figure 4.10. Here one uses the effect that the suction pressure of a jet compressor is influenced by the discharge pressure, when this discharge pressure is higher than the critical discharge pressure. The jet pump then operates on the right hand (declining) part of its characteristic that is shown in Figure 4.7. This kind of control is mainly used for ejectors with large suction capacity and large suction nozzles, because the alternative would be very big and expensive control valves in the suction line for controlling the suction pressure. However, it has to be checked that this kind of control is applicable, because the decline of the characteristic can be very steep, in particular for high compression ratios (ratio of discharge pressure/suction pressure), thus stable plant operation will not always be possible.

A nozzle needle can also influence the motive flow rate of a jet compressor. The installation of such a nozzle needle controlled ejector is shown in Figure 4.11. The needle is adjusted by a pneumatic or electric actuator, as used for standard control valves. When a nozzle needle control is used, the motive pressure is not influenced. In contrast to the throttle control, described above, the motive mass flow is reduced without reducing the motive pressure that is required for the compression. The efficiency of a nozzle needle controlled ejector for partial load is therefore higher than with a throttle control valve.

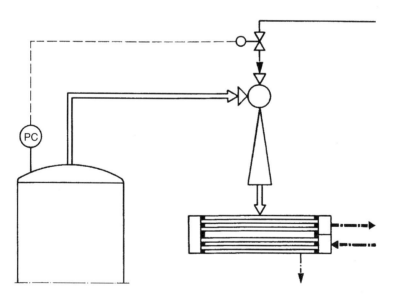

Figure 4.10 Suction pressure control by influencing the motive pressure.

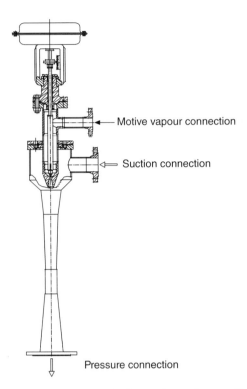

Figure 4.11 Assembly of a nozzle needle controlled ejector.

The adaptation of the suction capacity of an ejector can not only be achieved by influencing the motive fluid supply, but also by measures on the suction side. Here, the suction capacity can be influenced by a control valve in the suction line or by an additional ballast load (Figure 4.12).

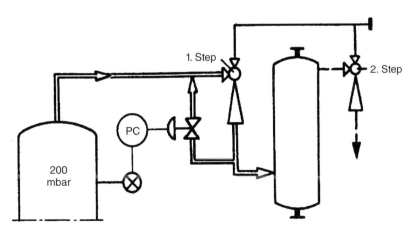

Figure 4.12 Vacuum control by a bypass (additional load on suction side).

If using a control with an additional ballast load it is normally reasonable to use a bypass from the outlet of the ejector back to the suction side and thus not give an additional external load to the system. For example, when pollutants are in the suction flow and ambient air is sucked in, this air will be saturate with the toxic components and carry them out and pollute the environment.

4.4
Multi-Stage Steam Jet Vacuum Pumps

The economically reasonable maximum compression ratio of a steam jet compressor is about 15–20. However the maximum discharge pressure can be less than 10, depending on the available motive steam pressure and the discharge pressure that has to be achieved. As steam jet vacuum pumps are used for suction pressures from 0.001 to 1000 mbar, with a main application range between 0.1 and 100 mbar, it is obvious that a multi-stage system has to be used in most cases. Only for suction pressures higher than 150 mbar are steam jet vacuum pumps normally build as single stage systems.

If possible, condensers are installed between the ejector stages to, as far as possible, condense the motive steam of the upstream ejector and condensable components of the suction flow. By so doing, the load on the downstream ejector stage is reduced, meaning that it can be smaller and with a lower motive steam consumption. If, for example a steam jet vacuum pump for a suction pressure of 1 mbar has to be designed, the lowest possible operation pressure for a condenser is determined. This minimum condensation pressure primarily depends on the cooling water temperature and the heating in the condenser. In the example, the cooling water has a maximum temperature of 25 °C. If this cooling water is heated by 10 degrees in the first inter-condenser, that is from 25 to 35 °C, the condensation temperature is higher than 35 °C. When using a mixing condenser a condensation temperature of 36 °C is sufficient. This gives a condensation pressure of 60 mbar (saturation pressure for 36 °C).

In this example (see also Figure 4.13), the suction flow has to be compressed from 1 to 60 mbar before it will be possible to condense the motive steam. A compression ratio of 60 : 1 cannot be handled in a single stage. Therefore, two jet compressor stages have to be connected in series and the second one of these two pre-stages or boosters has to handle both the suction flow and the motive flow of the first stage. The further compression from 60 mbar to atmospheric pressure is then done with two more ejector stages, however with a condenser after each stage. At pressures higher than 60 mbar, it is always possible to condense with cooling water of 25 °C.

Figure 4.13 shows the setup of the described four-stage steam jet vacuum pump with two inter-condensers and one after-condenser for a first stage suction pressure of 1 mbar.

Since the non-condensable gases (also called inert gas, e.g. air) that leave a condenser are saturated with water vapour, it is important that every ejector stage

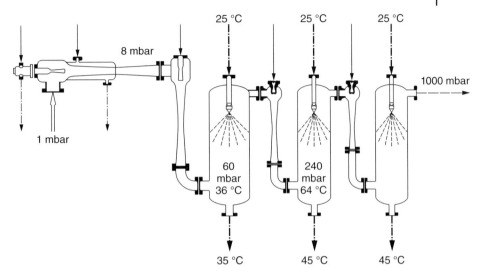

Figure 4.13 Four-stage steam jet vacuum pump with mixing condensers.

operating downstream of a condenser, can handle not only the non-condensable components of the suction flow, but also a water vapour portion. In vacuum, this portion can be quite high. More details for this can be found in the previous chapter about condensation.

As the specific volume of vapours and gases decrease with increasing pressure, the volume flow rate is reduced during the compression from 1 to 1000 mbar and the dimensions of the ejectors and condensers are also scaled down from stage to stage.

Surface condensers are used instead of mixing condensers, if the suction flow contains components that are not permitted to be mixed into the cooling water. A schematic representation of a four-stage steam jet vacuum pump with surface condensers is given in Figure 4.14.

For the design of multi-stage steam jet pumps it is important to consider a sufficient overlap, that is safety margin, between the discharge and suction conditions of successive ejector stages. For stages that work directly in series, that is one after the other without an intercondenser, the suction pressure and the suction flow rate of the downstream ejector is directly influenced by the outlet conditions of the previous stage. If a condenser is installed between two stages, the conditions are different. The vapour–liquid equivalent at the condenser outlet has to be calculated, with regard to the inert gas flow rate and the cooling water conditions. From this the necessary suction capacity of the downstream stage can be defined.

Figure 4.15 shows an example for a multi-stage steam jet vacuum pump with a material combination of porcelain and graphite. These materials are suitable if aggressive components are in the suction flow.

Frequently a combination of steam jet pumps and liquid ring pumps is installed. Here, the low suction pressure is achieved by one or several ejector stages, but

4 Steam Jet Vacuum Pumps

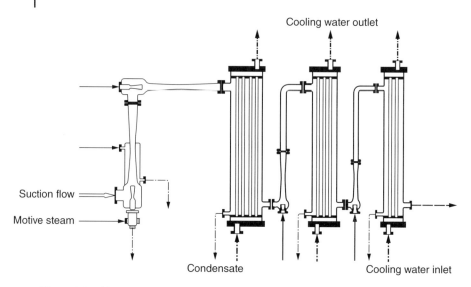

Figure 4.14 Four-stage steam jet vacuum pump with surface condensers.

Figure 4.15 Four-stage steam jet vacuum pump, ejector stages in porcelain, block condenser made from graphite.

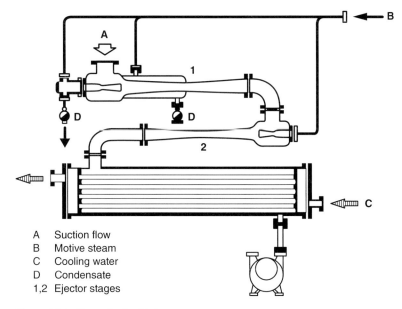

Figure 4.16 Steam jet liquid ring vacuum pump

A Suction flow
B Motive steam
C Cooling water
D Condensate
1,2 Ejector stages

a liquid ring pump is used to compress against atmosphere as the last stage (Figure 4.16).

4.5
Comparison of Steam, Air and Other Motive Media

If jet pumps are driven by steam, waste water is generated when the motive steam is condensed. In this condensate, residuals of product from the suction flow can often be found. Even though this waste water flow rate is small and the costs for handling the effluents are similarly small, it should be an aim to reduce the waste water flow as far as possible. When the ejector is driven by a product vapour, this problem can often be avoided. Normally the motive medium condensate can again be evaporated and reused as motive vapour (Figure 4.17).

A further field of application for product vapour driven ejectors are in processes where the intrusion of water has to be avoided at all costs. In addition, the use of motive media other than steam can also be energy saving.

The design of product vapour driven jet pumps needs a detailed analysis in order to find the best degree of efficiency. The main fluid dynamic processes in a vapour- or gas-driven ejector are firstly the expansion of the motive medium, then the mixing of motive and suction flows and finally the compression of the mixed flow to the required discharge pressure. These processes are influenced by the geometry of the ejector, the properties of the media employed and the prevalent pressures.

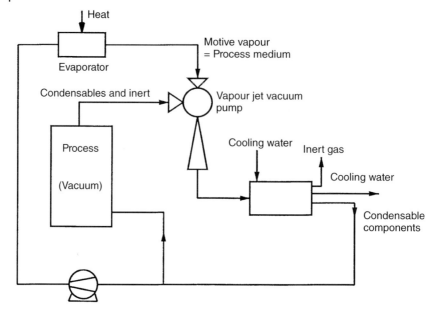

Figure 4.17 Schematic illustration of an effluent free vapour jet vacuum pump with surface condensation.

Using the calculation methods of one dimensional gas dynamic shows how different motive media behave during the expansion in the motive nozzle. Depending on the expansion ratio (E = motive vapour pressure/suction pressure), the Mach number at the motive nozzle outlet for some vapours is charted in Figure 4.18. It can be seen, that the velocity is in the supersonic range and therefore the corresponding effects (expansion, compression and shock waves) have to be considered.

The applicability of a substance to be used as a motive medium in a vapour jet vacuum pump is also considerably influenced by its vapour pressure diagram. It must be considered that the evaporation temperature is always below a possible

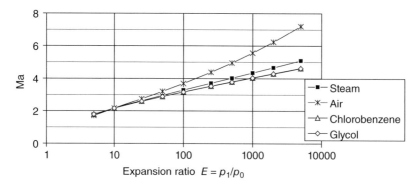

Figure 4.18 Mach number at motive nozzle outlet.

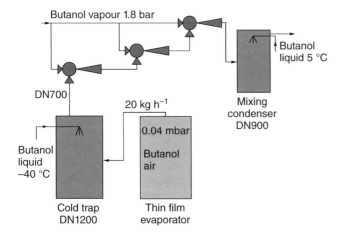

Figure 4.19 Example for a vapour jet vacuum pump with butanol vapour as motive medium.

pyrolysis temperature and that the condensation temperature is higher than the triple point.

The temperature of the available cooling water and the possible condensation pressure can preclude the use of a special motive medium. Also, the variation of the vapour pressure diagram by the addition of other components must be regarded. The example of the twin component system, ethylene glycol/water, makes this point more clearly. Even small parts of water lead to a distinct decrease of the necessary cooling liquid temperature. For example, at 2 mbar and a water content of 1% condensation is possible at 35 °C, but even at 3% water content the condensation temperature has decreased to about 15 °C.

Finally, Figure 4.19 shows the layout of a three-stage vacuum pump without waste water output. Here butanol is used as motive medium, as cooling liquid in the mixing condensers and also in the liquid ring pump which is not shown here.

Further Reading

Bartscher (1990) Aufbau und Anwendungsmöglichkeiten von Strahlpumpen. *Chemie-Technik*, **19** Nr. (2), 40–45.

BHRA (1984) Jet Pumps and Ejectors.

Grave, H. (2002) *Gas- und Dampf-Strahlpumpen*, Verlag und Bildarchiv W.H. Faragallah.

Jousten, K. (ed.) (2008) *Handbook of Vacuum Technology*, Wiley-VCH Verlag GmbH, pp. 375–388.

Martin, G.R. (1997) Understand real-world problems of vacuum ejector performance. *Hydrocarbon Processing*, **76** (11), 63–75.

Power, R.B. (1993) *Steam Jet Ejectors for the Process Industries*, McGraw-Hill.

Produktkatalog (2009), GEA Wiegand GmbH, Ettlingen.

5
Mechanical Vacuum Pumps
Wolfgang Jorisch

5.1
Introduction

In connection with the large-scale industrial development in the area of chemical process engineering, which took place during the first decades of the twentieth century, vacuum technology was needed for many process steps chiefly to perform here process steps at pressures below atmospheric pressure, thereby creating favourable conditions for the product.

At that time and still today, typical process steps involve distillation as well as drying – process steps which can be attributed to the area of thermal process engineering.

Compared to vacuum applications in other industrial areas, chemistry and pharmaceutical applications typically involve large quantities of condensable vapours which need to be removed utilising a vacuum on the side of the process which are then compressed to atmospheric pressure (see chapter 2 in this book as an example). Under atmospheric conditions or already during compression within the vacuum pump on the way to atmospheric pressure, the vapours being pumped out involving a vacuum may condense thereby forming liquids whereby this can only be avoided with difficulty, if necessary.

The most frequently utilised type of vacuum pump in these areas of application – even since the broad utilisation mechanical vacuum pumps – was and still is today in the area of industrial chemical process engineering, the so-called water ring pump or liquid ring vacuum pump. This type of vacuum pump is covered in-depth in chapter 3 of this book.

The liquid ring vacuum pump will show its weaknesses where processes involving relatively low operating pressures, which physically certainly can belong to the rough vacuum range but which owing to the vapour pressures of the commonly used operating agents for the liquid ring vacuum pump, can only be implemented with greater complexity (cooling of the operating agent to lower temperatures, use of pump systems and thus more involved pump combinations) (pressures down to 1 mbar and a few 10 mbar).

Vacuum Technology in the Chemical Industry, First Edition. Edited by Wolfgang Jorisch.
© 2015 Wiley-VCH Verlag GmbH & Co. KGaA. Published 2015 by Wiley-VCH Verlag GmbH & Co. KGaA.

Here up to approximately the 90s of the twentieth century, the oil-sealed rotary vane vacuum pump provided its contribution towards filling this gap.

The deciding factor for its successful deployment was, that it was possible, process engineering-wise and under practical everyday plant operating conditions, not to allow product vapours to condense in this oil-sealed rotary vane vacuum pump, but instead to let condensation take place only in a downstream exhaust gas condenser external to the vacuum pump.

Since in practice maintaining of these conditions was frequently found to be extremely difficult and since any early condensation within the rotary vane vacuum pump would result in considerable damage and thus high repair costs owing to the dilution of the oil, fresh-oil-lubricated rotary vane vacuum pump types were perfected for deployment in the chemical process industry.

In this type of pump, the avoidance of condensation is no longer an absolute requirement as for oil-lubricated pumps with circulatory lubrication, even though it is still desirable to prevent any condensation also in fresh-oil-lubricated rotary vane vacuum pumps (risk of diluting the fresh oil too much). In any case, condensate which forms here can be discharged, since the exhaust of these pumps is not located at the highest point as is the case for types with circulatory lubrication, but instead the exhaust is located at the lowest point.

The disadvantage involved in using these fresh-oil-lubricated vacuum pumps was the provision of fresh-oil logistics and waste oil disposal, which was cumbersome when operating many such vacuum pumps constantly.

Towards the end of the 80s of the twentieth century, besides having to solve process engineering problems (low operating pressures, avoidance of condensation and alternatives to liquid ring vacuum pumps) more and more questions related to the protection of the environment arose which now also had to be solved. If unsuccessful in closing the operating fluid circuit of a liquid ring vacuum pump, greater quantities of operating fluid will have to be constantly replaced and the plant will also have to solve a waste problem unless the produced condensates can be fed back into the process, whereby this would be not always be feasible, and if so only in special cases. These questions of environment protection finally were the incentive, besides the previously used alternative of using an oil-sealed vacuum pump, to investigate an entirely new approach and develop even more suitable vacuum pumps which preferably would do entirely without any operating agents and which, owing to their absence could not be contaminated and which therefore would not produce any additional waste.

Also more and more new requirements regarding the purity of the entire processing chain especially in connection with pharmaceutical products arose, because of which dispensing with any kind of operating agent in vacuum pumps was demanded. In pharmaceutical production facilities, an operating agent in a vacuum pump always implies a potential source of process contamination, for example simply by backstreaming of the operating agent in case of unintentional improper operation of the vacuum section of the plant or just by contaminating the operating fluid with microbiological activity thus incurring unwanted bacterial growth.

5.2 The Different Types of Mechanical Vacuum Pumps

Old and new requirements from the area of chemical process engineering regarding the process vacuum pumps described, resulted beginning at the end of the 80s of the twentieth century in considerable development efforts of all leading manufacturers of mechanical vacuum pumps so as to be able to offer process-capable dry compressing pumps for the area of chemical and pharmaceutical process engineering.

The following covers the operation of all these vacuum pumps and their utilisation especially in the chemical process industry.

5.2
The Different Types of Mechanical Vacuum Pumps

The vacuum processes which are utilised in the area of chemical engineering are generally performed in the range of one to several hundred millibars (Figure 5.1). An exception to this is the process of short path distillation, respectively molecular distillation, which operates at low pressures in the medium and the beginning high vacuum range and which here is separately covered in chapter 15 of this book.

Depicted in Figure 5.2 are the common operating pressure ranges of mechanical vacuum pumps.

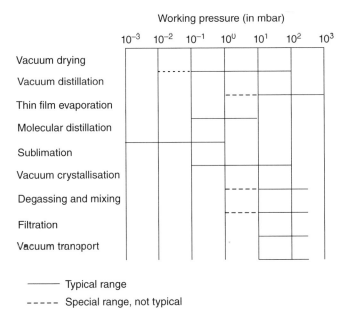

Figure 5.1 Typical processes working under vacuum in chemical engineering.

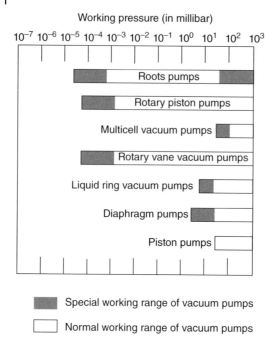

Figure 5.2 Common operating pressure ranges of mechanical vacuum pumps.

The following mechanical pump principles (positive displacement vacuum pumps) are utilised in the construction of vacuum pumps [1].

5.2.1
Reciprocating Piston Vacuum Pump

'A reciprocating piston vacuum pump (Figure 5.3) is an oscillating type of positive displacement vacuum pump in which the wall moving to and fro is a piston moving within a cylinder'. Reciprocating piston vacuum pumps are generally pumps of an older design whereby lubrication is effected either through circulatory lubrication or by way of fresh oil lubrication. Based on this operating principle also smaller dry compressing vacuum pumps have been developed.

5.2.2
Diaphragm Vacuum Pump

'A diaphragm vacuum pump is an oscillating type of positive displacement vacuum pump in which the wall moving to and fro is a diaphragm' (Figure 5.4). Today, diaphragm vacuum pumps have become a popular environment-friendly alternative to water jet vacuum pumps (kinetic vacuum pump) at low and medium pumping speeds especially in chemical laboratories and in connection with pilot

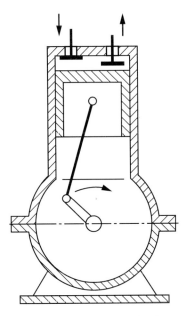

Figure 5.3 Working principle of a piston pump.

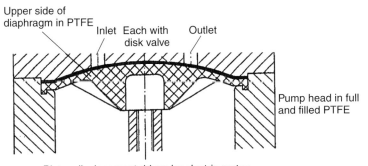

Figure 5.4 Principle of modern diaphragm vacuum pump.

plants. Here the parts in contact with the medium are commonly manufactured using perfluorinated polymer materials for the purpose of attaining a universal resistance against media.

5.2.3
Rotary Vane Vacuum Pump

'A rotary vane vacuum pump (Figure 5.5) is a rotating type of positive displacement vacuum pump in which an eccentrically suspended impeller slides tangentially past the inside wall of the stator (casing). Two or more vanes with the capability

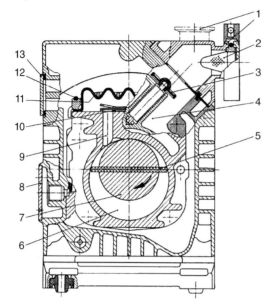

Figure 5.5 Sectional drawing of a rotary vane vacuum pump operating with two vanes. (With permission from Oerlikon Leybold Vacuum, Cologne, Germany.)

Main components:
1 = Suction port (changeable vertical or horizontal),
3 = Back pressure valve,
5 = Rotary piston with two vanes (other types can have three or more vanes),
6 = Suction room,
8 = Gas ballast,
10 = Exhaust valve, and
13 = Connection for accessories like exhaust oil demister (in some other types integrated in the pump itself)

of being able to move in impeller slots rest against the inside wall of the stator (the casing) dividing the pump chamber into chambers changing in volume'.

Rotary vane vacuum pumps are commonly provided with circulatory lubrication (chiefly in more physical or physical–industrial applications); in the chemical industry frequently with fresh oil lubrication and then of a multi-vane design. Rotary vane vacuum pumps of both types are characterised in that with these it is relatively simple to attain very low vacuum pressures.

5.2.4
Rotary Plunger Vacuum Pump

'A rotary plunger vacuum pump is a rotating type of positive displacement vacuum pump in which a piston moves eccentrically along the inside wall of the casing. A vane rigidly joined to the impeller divides during its rotation the pump chamber into two chambers changing in volume'. Figure 5.6 depicts the work cycle of a rotary piston vacuum pump. For the use of this type of pump in the area of chemical process engineering fundamentally the same statements apply as detailed below for oil-sealed rotary vane vacuum pumps; only designs with circulatory lubrication are known.

As for the rotary vane vacuum pump, condensation of pumped solvents in the vacuum pump resulting in a dilution of the lubricating oil can only be accepted to a very limited extent, preferably not at all.

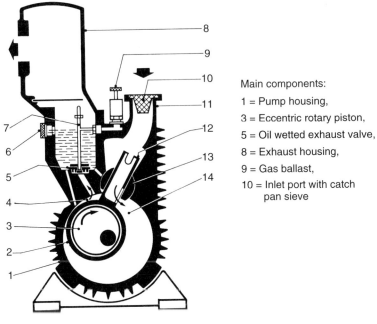

Figure 5.6 Working principle of a rotary piston vacuum pump. (With permission from Oerlikon Leybold Vacuum, Cologne, Germany.)

Main components:
1 = Pump housing,
3 = Eccentric rotary piston,
5 = Oil wetted exhaust valve,
8 = Exhaust housing,
9 = Gas ballast,
10 = Inlet port with catch pan sieve

5.2.5
Roots Vacuum Pump

'A Roots vacuum pump is a rotating type of positive displacement vacuum pump in which two impellers having the same cross-section, commonly exhibiting the shape of a figure-of-eight, revolve about each other in opposite directions, without touching each other and the casing wall. The two impellers are so synchronised that they move past each other with only a slight amount of play. No compression is effected within the pump chamber of the pump' (isochoric compression).

Depicted in Figure 5.7 is a cross-sectional view through a Roots vacuum pump. Beneficial here to chemistry applications is the additional separation of the pump chamber with respect to the gear chamber, in which also a vacuum is maintained, this separation being provided through a seal gas facility. This seal gas which is fed in from the outside between the piston rings of the labyrinth seal ensures that there always exists an albeit very slight inert gas flow away from the gear in the direction of the pump chamber, thereby preventing the ingress of corrosive vapours or gases, for example into the gear chamber.

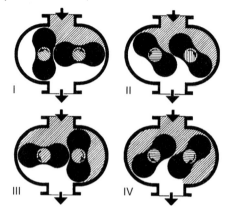

I = Gas arrives at suction port,
II = Rotors are turning, gas begins to be enclosed,
III = Isochoric enclosure of gas is continued, and
IV = Gas begins to leave the pump at exhaust without an internal compression.

Figure 5.7 Lobe type (Roots) vacuum pump.

5.2.6
Dry Compressing Vacuum Pump

Dry compressing vacuum pumps have been developed for deployment in the chemical industry. They are of a multistage design based on the so-called claw principle, as a multistage Roots vacuum pump or nowadays as screw type vacuum pumps, which is the most up-to-date type of compression in dry vacuum pumps for chemical engineering. All these dry pumps have in common that the active pumping chamber is free of any sealing or lubricating agents. Details of these types of vacuum pumps are reported later in this chapter.

5.3
When Using Various Vacuum Pump Designs in the Chemical or Pharmaceutical Process Industry, the Following Must Be Observed

5.3.1
Circulatory-Lubricated Rotary Vane and Rotary Plunger Vacuum Pumps

As opposed to liquid ring vacuum pumps, it needs to be ensured under all circumstances in the case of this type of vacuum pump that all sucked-in vapours can exit the vacuum pump without condensation, that is also by way of vapour in the gaseous state.

If this is not accomplished, condensation occurs in oil-sealed rotary vane or rotary plunger vacuum pumps, resulting in a severely impaired ultimate pressure and pumping speed and, moreover, these vacuum pumps may suffer significant damage.

Substances incapable of dissolving in oil, like water vapour, form when condensing with the oil, an emulsion. Oil-soluble substances like condensed hydrocarbon vapours much dilute the oil with correspondingly negative effects on attainable ultimate pressure and pumping speed. Moreover, in both cases the lubricating effect of the oil, a task which the oil must fulfil in these types of vacuum pump, is much impaired. Thus the possibility of damage due to wear affecting the pump chamber or the vanes with their guiding slots in the impellers, cannot be excluded.

In order to prevent taken-in vapours from condensing during the compression phase from vacuum to atmospheric pressure, the rotary vane and rotary plunger vacuum pumps are set up for a 'vapour tolerance' which is as high as possible. The term 'vapour tolerance' denotes the maximum permissible partial pressure of a vapour on the intake side of the pump which, when complied with, respectively when not exceeded, will not give rise to any condensation in the vacuum pump even when compressed to atmospheric pressure. A very important influencing quantity which has an impact on the level of vapour tolerance, is the operating temperature of the vacuum pump, since owing to a good thermal design on a temperature level which is as high as possible but also as uniform as possible, any condensation in the pump becomes more and more difficult. This measure is naturally limited through the then ever decreasing service life of the oil (thermal stress on the oil).

The approach of increasing the temperature level within the vacuum pump is aimed at exceeding the saturation vapour pressure of the pumped vapour at all times also during the compression phase. If water vapour, for example is sucked in from a vacuum and when maintaining the temperature level of the entire vacuum pump at over 100 °C, one can speak of an unlimited degree of water vapour tolerance provided the pressure on the exhaust side does not exceed atmospheric pressure. The solution of ensuring everywhere on the pump, also at the coldest spots of a pump, temperatures at such a level is not particularly elegant, since besides the use of highly temperature resistant types of oil also the entire rotary vane pump with its close tolerance slots and its elastomer seals must be suited for such high temperatures. Moreover, organic substances which are also being pumped may carbonise in a very hot vacuum pump thereby forming unwanted particles in the oil of the vacuum pump.

The most effective method of increasing the level of vapour tolerance in an oil-lubricated rotary vane vacuum pump is that of providing a gas ballast, invented by Gaede. In the case of operation with a gas ballast, it is in connection with mechanical vacuum pumps not necessary to aim for a very high temperature level. Even so, an operating temperature level of 60–80 °C and a temperature spread which is as uniform as possible should preferably be maintained so as not to provoke condensation at any point.

Operation of the gas ballast is depicted in Figure 5.8.

As soon as the intake process has been completed and before the compression process begins, a precisely regulated 'quantity of inert gas' as the non-condensable share is sucked in (*step b2*). Thus the 'total pressure' in the pump chamber is increased without moving the piston, that is without compression, opposite to the situation in *step a2*.

After having increased the total pressure in the pump chamber by admitting the ballast gas (*see step b2, not so in step a2*), only a rather small compression ratio in

5 Mechanical Vacuum Pumps

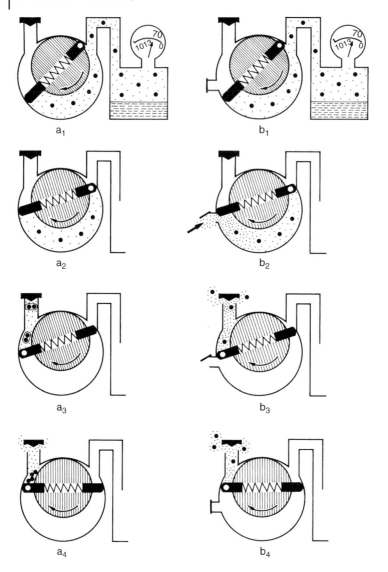

Figure 5.8 Function of a rotary vane vacuum pump without (series a1–a4) and with (series b1–b4) gas ballast.

the vacuum pump is necessary in order to attain the atmospheric pressure level needed for ejection (common compression ratio 1 : 10). During this compression process the partial vapour pressure in the vacuum pump is in *steps b3 and b4* only increased by this factor, not so in *steps a3 and a4*, where the pressure of the vapour is compressed up to atmospheric values with the danger of internal condensation. Thus aided by the gas ballast the possibility of attaining the saturation vapour pressure and thus condensation within the pump chamber is reduced, that is the desired effect is precisely attained (*steps b3 and b4*).

As already mentioned, the conditions with and without gas ballast are depicted in Figure 5.8.

In accordance with DIN 28 426 Part 1, the vapour tolerance of a mechanical vacuum pump which shall pump the taken in vapours without condensing during the compression phase from vacuum to atmospheric pressure, is defined as follows:

$$P_{D_V} = \frac{B}{S} \frac{1333 \; (P_D - P_A)}{1333 - P_D} \; \text{mbar} \tag{5.1}$$

P_{D_V} = vapour tolerance (mbar); B = gas ballast quantity (m³ h⁻¹); S = pumping speed of the vacuum pump (m³ h⁻¹) and P_D = vapour pressure of the pumped product at operating temperature of the vacuum pump (may, owing to solubilities in the pump's operating medium be subject to a lowering of vapour pressure).

P_A = partial pressure of the pumped vapour in the gas ballast (when charging the gas ballast with this vapour in case of a gas ballast return from a downstream emission condenser, for example);

1333 = exhaust backpressure in millibar (abs.) on the exhaust side of the vacuum pump.

From Eq. (5.1) it is apparent which parameter setting ensures a high and thus desired vapour tolerance P_{D_V}

- a large gas ballast quantity increases the vapour tolerance. However, the gas ballast quantity can be only increased within certain limits. Maximum 5–10% of the available pumping speed is commonly used for the gas ballast. Facilities like the gas ballast are utilised also in connection with modern dry compressing pumps so as to prevent unwanted condensation effects here too. Inasmuch the aspects discussed here can also apply to the utilisation of modern dry compressing vacuum pumps, even though the design of such a modern dry compressing pump should be such that condensing solvent vapours can be taken in and discharged again without damaging the pump.
- A high vapour pressure P_D at operating temperature will result in a high vapour tolerance. Thus high vapour tolerances can be expected in the case of substances which boil easily, especially when they additionally exhibit a poor solubility in the pump's oil (polar substances). Examples for such substances are acetone or methanol. From these relationships it additionally becomes apparent that an adequate vapour tolerance may in any case only be expected from a rotary vane or rotary plunger pump which has been allowed to warm up (only then will the vapour pressure be high). A pre-operation phase before starting the process will in any case be necessary to warm up these vacuum pumps when condensable vapours must be pumped.
- A backpressure on the exhaust side which is too high will result in an increased tendency of promoting condensation of the pumped vapours. In Eq. (5.1), an already increased exhaust backpressure of 1333 mbar with respect to atmospheric pressure has been taken into account. This is expedient, in view of washing units or long exhaust gas collection systems installed in the exhaust

Table 5.1 Vapour tolerances of a modern rotary vane pump with a water vapour tolerance of 40 mbar.

Substance	Maximum permissible vapour pressure of the pure vapour on the intake side of the vacuum pump (mbar)
Acetone	1013
Dichloromethane	1013
Ethanol	260
Hexane	135
Methanol	1013
Propanol-2	465
Toluene	19
Water	40

line which give rise to an increased backpressure. In any case and with respect to a high vapour tolerance, the backpressure on the exhaust side should be kept as low as possible.
- The vapour tolerance is also negatively influenced by a too high partial pressure of the pumped vapour in the gas ballast. This means that a fresh gas ballast (without any gas return from the exhaust side) will here be most beneficial, but will under certain circumstances also increase the exhaust emissions.

Given in Table 5.1 are some typical vapour tolerance values for a rotary vane pump with circulatory oil lubrication.

Generally a value of 40–60 mbar maximum is aimed at for the water vapour tolerance limit. This results from the fact that a condenser operated upstream of the vacuum pump and with normal circulating cooling water will reduce the partial pressure of the water vapour down to 40 mbar and up to 60 mbar.

5.3.2
Fresh-Oil-Lubricated Rotary Vane Vacuum Pumps

Fresh-oil-lubricated ('once through') rotary vane vacuum pumps are generally vacuum pumps of the multicell type. As to their design they differ only insignificantly from the rotary vane pumps detailed above, since operating principle and compression process are very similar (Figures 5.9 and 5.10).

Basically, multicell pumps are equipped with more vanes per impeller (commonly six vanes). The vanes of a multicell pump can only be moved by the centrifugal force and pressed against the inside wall of the pump chamber.

When oil-lubricated rotary vane vacuum pumps are used in connection with a chemistry application, then usually fresh-oil-lubricated vacuum pumps will be chosen. The oil which in all rotary vane vacuum pumps serves the purpose of inner sealing and lubrication is constantly fed into the pump chamber and thus passes together with the gas, respectively vapour flow through the pump.

5.3 When Using Various Vacuum Pump Designs in the Chemical

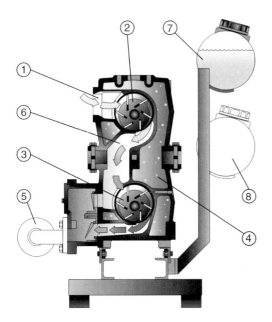

Main components:
1 = Suction portx,
2 = Rotor first stage,
3 = Rotor second stage
 (both fitted with 6 vanes),
4 = Water cooling,
5 = Exhaust,
6 = Gas transfer first to second stage,
7 = Fresh oil tank, and
8 = Rinsing oil tank (option).

Figure 5.9 Sectional drawing of a fresh-oil-lubricated rotary vane vacuum pump. (With permission from Dr.Ing-K. Busch GmbH, Maulburg, Germany.)

Figure 5.10 Picture of a ready to use fresh-oil rotary vane vacuum pump, direction of gas transport vertical (suction port up and exhaust down), including fresh oil tank and water/air cooler. (With permission from Dr.-Ing. K. Busch GmbH, Maulburg, Germany.)

The oil quantities within the pump chamber are relatively small, since the use of too much oil is prohibitive in view of the then occurring waste oil quantities. Because of the thus unavoidably restricted possibility of dissipating the compression heat, the active pumping section is cooled by means of a double water jacket. Fresh-oil-lubricated vacuum pumps which are intended for deployment in chemical production lines are manufactured with pumping speeds ranging from 50 to approximately 700 m^3 h^{-1}. The ultimate pressure attainable by these two-stage pumps reaches values down to 0.5 mbar (abs.)

As in these vacuum pumps the oil passes only once through the pump, the purity of the separated oil is of far less significance compared to rotary vane pumps with circulatory lubrication. Even so, also in the case of fresh-oil-lubricated pumps, the aspect of vapour tolerance must be taken into account because here too the oil may suffer from dilution or the oil film in the pump chamber may be even washed away with the result of very rapid and considerable wear within the pump chamber giving rise to high repair costs.

Due to the unavoidable oil consumption of a fresh-oil-lubricated rotary vane pump and many operating parameters which need to be complied in connection with chemistry applications like

- maintaining of a pre-operation phase for warming up,
- avoidance of condensation within the pump chamber and
- maintaining of a post-operation phase at the end of the process,

the range of applications for both the circulatory-lubricated and also the fresh-oil-lubricated rotary vane pumps in chemical production facilities is much dependent on the possibilities of process control or the application in general. Inasmuch especially when using these types of vacuum pump, professional planning is of paramount importance for success.

5.3.3
Dry, Respectively Oil-Free Compressing Vacuum Pumps

In accordance with DIN 28 400 (Part 2) a dry compressing vacuum pump is a positive displacement vacuum pump which operates without an oil seal (as the liquid seal). A diaphragm vacuum pump is a dry compressing pump, a type of pump which chiefly has its role in connection with laboratory applications. Here the diaphragm vacuum pump replaces the universal laboratory water jet pumps, the water consumption of which and thus the high operating costs due to the generation of contaminated waste water are factors which are not desired.

The diaphragm vacuum pump which in connection with chemical production facilities is practically of no great significance is not covered here. Information on this kind of pump can be found in [2].

For the same reasons on which the move in the chemistry laboratory to dry compressing vacuum pumps is based, dry compressing pumps have also been developed for chemical production facilities. The different pumps operate according to different principles. They are designed as backing pumps capable of compressing directly from vacuum against atmospheric pressure.

Chiefly the following fundamentally known vacuum pump principles have been relied on in the development of oil-free compressing pumps:

- Roots pumps (Roots blowers) which may be used as a dry compressing backing pumps alone, also for pump combinations with all kinds of 'backing pumps';
- so-called claw compressors, respectively their combination with Roots blowers;
- vacuum pumps, the operation of which is based on the screw principle.

Dry compressing rotary vane vacuum pumps based on multi-vane designs with self-lubricating vane materials are hardly used. This dry compressing type of vacuum pump has been found to be too prone to wear if used in chemistry applications.

Chemical process engineering demands of vacuum pumps a high level of operational reliability. Also no waste, like waste oil or waste water shall be produced as far as possible. The requirement as to the avoidance of additional waste quantities is optimally fulfilled by dry compressing vacuum pumps. However, the question as to their operational and process reliability arises. Here the benchmark for the intended level of process reliability should always be the liquid ring vacuum pump.

5.3.4
Roots Vacuum Pumps

Initially Roots vacuum pumps served the purpose of assembling pump combinations, so-called pump systems. The main reason for this is the relatively high pumping speed they offer in connection with their small size, however, usually with a rather small pressure difference which can be overcome.

5.3.4.1 Operating Principle of Roots Vacuum Pumps

The Roots principle is based on two impellers exhibiting the shape of a figure-of-eight, which revolve without making contact with each other and the casing. The slots between the impellers and the casing amount to only a few hundred micrometres, (see Figure 5.7) depending on the size of the pump. The gear wheels which ensure synchronous running of the impellers are accommodated in chambers (oil-lubricated gear chambers) on the face side.

The pressure difference which can be overcome by Roots vacuum pumps in the pump chambers in which the pumped gas is not compressed, is limited. Their great advantage is based on the extremely favourable specific construction volume (construction volume per cubic metre of pumping speed). Roots vacuum pumps offering pumping speeds of over $10\,000\,m^3\,h^{-1}$ are being manufactured.

Owing to this characteristic, Roots vacuum pumps are chiefly employed in multistage pump systems for pre-compressing gases and vapours. The compression of the pumped gas or vapour here does not occur within the pump but outside in the connecting piping running to the so-called 'backing pump' which is installed by way of a vacuum pump of lower pumping speed downstream of the Roots vacuum pump. This backing pump may be an oil-sealed or other dry compressing vacuum pump, a liquid ring pump or a further smaller Roots pump, for example. This downstream pump will generally compress the gases to atmospheric pressure. Such pump combinations are called vacuum pump systems.

5.3.4.2 Roots Vacuum Pumps with Bypass Valve, Respectively with Frequency Controlled Motor

In the case of standard Roots pumps, measures must be introduced which ensure that the maximum permissible pressure differences due to restrictions imposed by the pumping principle itself between the intake and the exhaust port are not exceeded. In the simplest case this is achieved through the use of a pressure switch which switches the Roots vacuum pump on or off depending on the intake pressure. A Roots vacuum pump with integrated bypass, respectively differential pressure valve offers the capability to open the valve fully or partially, depending on the prevailing pressure difference between the intake and the exhaust side, respectively, and when the permissible pressure difference is no longer exceeded the bypass valve closes completely (Figure 5.11). Thus the Roots vacuum pump may already be switched on at atmospheric pressure without exceeding the permissible pressure difference. The Roots vacuum pump is no longer fully switched on as soon as a certain pressure is attained. Instead the pumping speed of the pump system is gradually increased.

However, in chemistry applications spring-loaded valves have been found to be at least failure prone. When pumping substances which cause sticking of such a bypass valve, its function is always endangered. There exists the considerable risk that in the event of a valve function failure, the Roots pump is damaged since under these conditions the permissible pressure difference can be exceeded.

Figure 5.11 Roots vacuum pump with internal bypass valve.

A more suitable and modern approach in connection with chemistry applications is that of smoothly 'cutting-in' a Roots vacuum pump in to a process without relying on a bypass valve but instead by using a frequency and thus speed controlled motor. Now the Roots vacuum pump will no longer start to operate at a certain operating pressure at full speed. Under certain circumstances it may be started even at atmospheric pressure, but under speed, respectively frequency control, and at a lower rotational speed and thus a low theoretical volumetric speed. The characteristic is the same as for a Roots vacuum pump equipped with a bypass valve, except that now the valve can no longer fail because it is missing.

The lower the intake pressure drops in the course of the process, the faster the Roots vacuum pump may revolve. This operating approach makes sense since at lower pressures the same gas quantities take up an ever increasing volume. Thus at low pressures generally also a higher pumping speed is needed.

Figure 5.12 depicts the smooth increase in the total pumping speed of a pump combination with a Roots vacuum pump with bypass valve, respectively without this valve but with frequency control.

5.3.4.3 Compression of Roots Vacuum Pumps

The choice regarding the size of the backing vacuum pump for a corresponding Roots vacuum pump, respectively the pumping speed of the backing pump in a combination, also expressed as grading of the pumps, depends on the desired operating pressure, respectively operating pressure range [3]. Criteria like the maximum permissible pressure difference of the Roots pump as well as the desired effective pumping speed of the combination need to be taken into account here. From the rotational speed n and four times the volume which is sealed off in the T position of the impellers by one of the two impellers (pump chamber volume V_s), the theoretical pumping speed S_{th} of a Roots vacuum pump

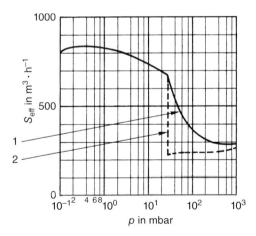

Figure 5.12 Suction speed diagram of a pump combination (fore vacuum pump + roots pump, roots pump without (2) and with (1) bypass valve respectively frequency controller.

is calculated as follows:

$$S_{th} = n \cdot 4 \cdot V_s$$

The quantity of gas Q_{eff} effectively pumped by a Roots pump vacuum pump is calculated from the theoretically pumped quantity of gas Q_{th} and the internal backflow Q_{IR} (as lost quantity) according to

$$Q_{eff} = Q_{th} - Q_{IR}$$

The following applies to the theoretically pumped gas quantity:

$$Q_{th} = p_a \cdot S_{th}$$

where p_a is the intake pressure at the intake port of the Roots pump and S_{th} is the theoretical pumping speed.

For the quantity of gas flowing back within the pump, that is the pumping loss, the following can be assumed:

$$Q_{IR} = p_v \cdot S_{IR}$$

where p_V is the forevacuum or backing pressure, that is the pressure on the exhaust side of the Roots vacuum pump and S_{IR} is a fictitious loss in pumping speed by way of a backflow.

The volumetric efficiency η of a Roots vacuum pump is calculated as

$$\eta = \frac{Q_{eff}}{Q_{th}}$$

When inserting the expressions for Q_{eff}, Q_{IR}, Q_{th} the following equation for the efficiency η is obtained:

$$\eta = 1 - \frac{p_v}{p_a} \frac{S_{IR}}{S_{th}}$$

When using the designation k for the amount of compression present between the intake port and the exhaust port of a Roots vacuum pump, then the following applies:

$$k = \frac{p_v}{p_a} \text{ respectively } \eta = 1 - k \frac{S_{IR}}{S_{th}}$$

The maximum compression of a Roots vacuum pump, a characteristic quantity for any Roots vacuum pump, is obtained in accordance with DIN 28 426 Part 2 at 'zero throughput' and is designated as k_0, where $\eta = 0$.

The maximum compression at zero throughput is commonly presented by way of a diagram versus the backing pressure p_v (Figure 5.13).

For the efficiency of a Roots pump, generally the following relationship applies:

$$\eta = 1 - \frac{k}{k_0}$$

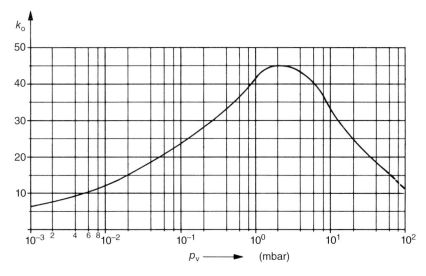

Figure 5.13 Maximum compression at '0-gas-transport' (k_0) of a roots pump depending on the different back pressures.

5.3.4.4 Dimensioning Combinations of Roots Vacuum Pumps with Backing Pumps

The effective pumping speed of a pump combination depends on the backing pressure p_V and thus on the pumping speed characteristic of the backing pump. The effective pumping speed of the pump combination is calculated as follows:

- From the pumping speed of the backing pump, a pumping speed S_V is determined for a specific backing pressure p_V.
- The intended combination of Roots vacuum pump (S_{th}) and backing pump (S_V) defines the theoretical grading

$$k_{th} = \frac{S_{th}}{S_V}$$

- Through the equation of continuity: $S_V \cdot p_v = S_{eff} \cdot p_a = \eta \cdot S_{th} \cdot p_a$ and through the equation for the efficiency

$$\eta = \frac{k_0}{k_0 + k_{th}}$$

one determines for the selected backing pressure p_V the corresponding maximum compression at 'zero throughput' k_0, calculates the efficiency η and finds from $S_{eff} = \eta \cdot S_{th}$ the effective pumping speed of the pump combination.

With the aid of a calculation programme it is thus possible to very quickly create pumping speed curves for pump combinations.

An important role in connection with selecting the pump combination is played by the maximum permissible pressure difference between the intake port and the exhaust port of the Roots vacuum pump, a value which commonly will be stated in the catalogues of the manufacturers.

A maximum pressure difference frequently found is the value of 80 mbar. An example shall serve the purpose of verifying which influence this maximum pressure difference has on dimensioning pump systems:

- The theoretical Roots pump pumping speed S_{th} is taken to be 1000 m³ h⁻¹.
- A combination with a backing pump is planned where the pumping speed of the backing pump S_V amounts to 100 m³ h⁻¹.
- The intake pressure upstream of the Roots pump p_A is taken to be 20 mbar.

Calculation Process There exists a theoretical grading between the Roots pump and the backing pump of 1 : 10.

The result of this is, that the Roots pump needs to provide a compression of 1 : 10 maximum, that is from 20 to 200 mbar. The therefrom derived pressure difference of 180 mbar is, with respect to the permissible pressure difference of 80 mbar maximum too high, so that this combination is not permissible.

When combining the same Roots vacuum pump with a backing pump providing a pumping speed of 250 m³ h⁻¹, there is no problem. The theoretical grading amounts to 1 : 4, that is the maximum compression across the Roots pump is taken to be from 20 to 80 mbar. The therefrom resulting pressure difference of 60 mbar is permissible.

But also the combination with a backing pump offering a pumping speed of 100 m³ h⁻¹ is possible when a different intake pressure p_A is demanded.

When considering the aforementioned combination (grading 1 : 10) now at an intake pressure upstream of the Roots vacuum pump of 5 mbar, the requirements regarding the maximum pressure difference are fulfilled.

At a grading of 1 : 10 and at an intake pressure of 5 mbar, a maximum pressure of 50 mbar will have to be assumed on the delivery side of the Roots vacuum pump. The pressure difference will now even in the most extreme case amount to only 45 mbar and is thus within the permissible range.

In order to be able to completely describe a pump combination as to its operating data, all that is now lacking is the calculation of the maximum cut-in pressure for the Roots vacuum pump. Meant here is the conventional mode of operating the Roots pump at a full rotational speed, without bypass valve and a programmed fixed cut-in pressure (through a pressure switch or through a pressure measurement arrangement from the side of the plant, for example).

This cut-in pressure P_e is calculated as follows:

$$P_e = \frac{\Delta P_{max}}{k_{eff} - 1} \text{ with } k_{eff} = \frac{S_{eff}(\text{Roots pump})}{S_{eff}(\text{Backing pump})} \text{ and}$$

$$k_{eff} \approx k_{th} = \frac{S_{th}(\text{Roots pump})}{S_{th}(\text{Backing pump})}$$

When running the calculation for the described combination of a Roots pump offering a theoretical pumping speed of 1000 m³ h⁻¹ and a backing pump with a pumping speed of 250 m³ h⁻¹ and a maximum permissible pressure difference across the Roots pump of 80 mbar then one obtains

$$k_{th} = \frac{1000 \text{ m}^3 \text{ h}^{-1}}{250 \text{ m}^3 \text{ h}^{-1}} = 4$$

and for the cut-in pressure of the Roots vacuum pump

$$P_e = \frac{80 \text{ mbar}}{4-1} = 26.6 \text{ mbar}$$

5.3.4.5 Power Requirement of a Roots Vacuum Pump

As already discussed, compression within a Roots vacuum pump being an outer compression is isochoric.

The following applies to the work on compression:

$$N_{compr.} = S_{th} \cdot \Delta P$$

In addition, also the mechanical power loss N_V of the machine needs to be provided:

$$N_{tot} = N_{compr.} + \sum N_V$$

As to the work on compression which needs to be provided, the following approximation applies:

$$N = \frac{S_{th} \cdot \Delta P}{36000} \text{ kW}$$

with S_{th} in m³ h⁻¹ and ΔP in mbar

Thus the rating for the drive motor for a given size of Roots vacuum pump depends on the permissible maximum pressure difference and the amount of mechanical power loss depending on the design in which also the amount of power required for starting up the pump must be taken into account.

Examples for the power uptake of Roots vacuum pumps, dependent on the pressure difference, are given in Table 5.2 for a pump with a theoretical pumping speed (S_{th}) of 1000 m³ h⁻¹:

5.3.4.6 Roots Vacuum Pumps with Pre-Admission Cooling Facility

The operating principle of a Roots vacuum pump with a pre-admission cooling facility is the same as for the pumps described above, but with the following difference.

The pumping process is normally completed after opening the pump chamber against the exhaust port of a Roots vacuum pump.

Table 5.2 Power requirement of a Roots vacuum pump as a function of the pressure difference.

p_a (mbar)	p_V (mbar)	Δp (mbar)	N_{el} (kW)
0.5	5	4.5	1.15
5	25	20	1.67
50	100	50	2.6

However, in the case of a vacuum pump with a pre-admission cooling facility, compressed and cooled down gas, cooled by means of a downstream gas cooler is admitted through the pre-admission channel into the pump chamber before the pump chamber is opened towards the exhaust port thereby ensuring direct cooling of the gas.

Thereafter the impeller, by continuing its motion, ejects the gas towards the exhaust port. By means of direct gas cooling, this type of Roots vacuum pump is capable of better dissipating the compression heat, so that a significantly higher pressure difference compared to Roots pumps without pre-admission cooling can be attained. Of disadvantage here are the relatively large heat exchanging surfaces needed for effective cooling of the gas in the vacuum range and the loud noise produced, due to the pumping principle, caused by the oscillating gas column in the gas cooler which acts as a resonator. Usually when deploying such Roots vacuum pumps, silencing measures like exhaust silencers and silencing covers are necessary. Figure 5.14 depicts the operating principle.

5.3.5
Dry Compressing Vacuum Pumps for Chemistry Applications

5.3.5.1   Dry Compressing, Three-Stage Roots Vacuum Pump with Exhaust, Respectively Non-Return Valves between the Stages

It is quite obvious to utilise the known principle of the Roots vacuum pump also for dry compressing machines.

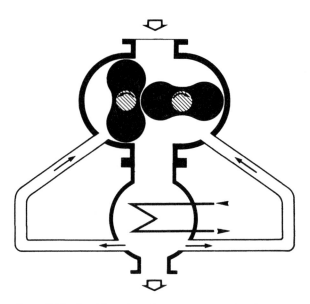

Figure 5.14 Special version of a roots pump with exhaust gas cooler to overcome a higher compression ratio (up to some 100 mbar).

These Roots vacuum pump combinations serve not only as boosters within the low pressure range where only a small pressure difference needs to be overcome, but also serve in multistage arrangements as a complete pumping unit, compressing from a few millibar abs. directly to atmospheric pressure.

The trade name of this water jacket cooled three-stage roots pump having internal backflow valves between each roots stage was 'Inovac'. The function can be described in short words as follows.

The Inovac operates with three stages which are based on series connected Roots vacuum pump stages. In order to prevent heated gas from back flowing into the respective compression stage, exhaust valves are a component of each stage. In order to dissipate the compression heat, which is a fundamental problem for all dry compressing vacuum pumps, a cooled water jacket is provided. This dry compressing chemistry vacuum pump attains through its three stages pressures down to under 1 mbar. The shaft seals for the oil-covered gear chamber consist of shaft sealing rings made of a material which is chemically resistant as far as possible. This type of pump is in the meantime no longer being developed further by the manufacturer since over time apparently difficulties occurred regarding a sufficiently stable design for the necessary intermediate valves, the operation of which needed to follow each work cycle. Generally, valves in vacuum pumps, regardless of type, serve the purpose of economically operating the vacuum pump but they always represent the most sensitive component within a vacuum pump, since the stresses to which the valves are exposed are unavoidably high.

From the presented design of a dry compressing pump the idea of the design engineers to simplify, respectively facilitate discharging of condensate produced during compression with the aid of the vertical construction, becomes readily apparent. Thus, similar to the conditions within liquid ring vacuum pumps it was for the first time possible with this dry compressing pump to allow the presence of condensate, provided such condensate would not be of a corrosive nature since it would be discharged and would not remain within the vacuum pump.

5.3.5.2 Claw Vacuum Pumps (Northey Principle)

Claw Vacuum Pump with Inner Compression So-called claw vacuum pumps have been developed both with, and also without inner compression of the taken in vapours, respectively gases. Such pumps with inner compression were and are being manufactured of the three stage type for chemistry applications and attain pressures below 1 mbar absolute.

Figure 5.15 depicts a sectional view of such a claw vacuum pump with inner compression and has the name EDP 250.

This claw vacuum pump is, just like the dry compressing roots combination described in the chapter before, cooled by means of a water jacket. The cooling water is recirculated and is cooled down by a heat exchanger, using the process cooling water through indirect heat exchange. The advantage is, that under these conditions the sometimes relative dirty process cooling water does not pass the cooling channels of the pump themselves, it just passes the primary side of the heat

5 Mechanical Vacuum Pumps

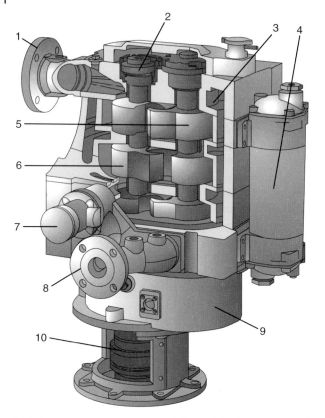

1. Inlet flange
2. Upper bearing in removable cartridge
3. Integral water cooling
4. Heat exchanger for cooling water circuit
5. Three stage claw mechanism
6. Reversed claws for shortest gas path
7. Blow-off valve for low power consumption
8. Outlet flange
9. Gear box
10. Clutch assembly

Dry pump model EDP250

Figure 5.15 Sectional drawing of the dry claw-type pump EDP 250. (With permission from Edwards Vacuum, Crawley, UK.)

exchanger. The shaft seals by way of shaft sealing rings are protected by a seal gas system against effects from the process and against contamination. By means of an overflow valve, any overcompression in the third pumping stage is prevented in the case of intake pressures which are too high. The overflow valve opens and guides the taken in process gases directly to the pump's exhaust. A direct and vertical pumping direction supports the ejection of any liquid shares possibly coming from the process. When stressing the vacuum pump with too much liquid, then a torque limiter will ensure that the electric drive unit is disconnected so that the pump mechanism is basically protected against suffering damage in the case entering liquid phases have a much higher viscosity compared to vapours and gases. Through

Figure 5.16 Claw type pump EDP 250 fitted with flame arrestors on suction and exhaust side, able to pump of ignitable mixtures of organic vapours in air. (With permission from Edwards Vacuuum, Crawley, UK.)

this pump, a quite compact dry compressing chemistry vacuum pump was developed which is capable of covering the entire rough vacuum range (Figure 5.16).

Claw Vacuum Pump Without Inner Compression This claw vacuum pump, specially developed for the area of chemical process engineering, is of a two-stage design and attains an ultimate pressure of below 10 mbar. It was marketed as a dry running alternative to the liquid ring vacuum pump. The pumped gases pass from top to bottom through the two vertically arranged stages, an arrangement which very much facilitates pumping of condensate or purging fluid which can become necessary from time to time (Figure 5.17).

Here too, the pump casing is cooled by a water jacket, however, in the case of this vacuum pump a further cooling component, that of feeding in cold gas is added (Figure 5.18).

The gases and vapours taken in from the vacuum are compressed in this type of vacuum pump not by reducing the volume of the pump chamber, but instead by venting with cold gas. Figure 5.19 depicts the pumping process.

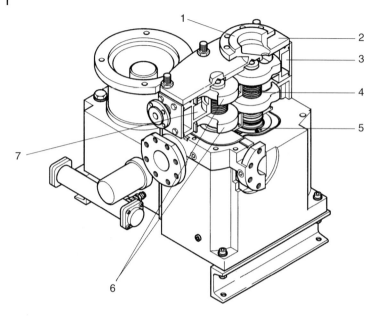

Figure 5.17 Two-stage claw-type pump without inner compression.

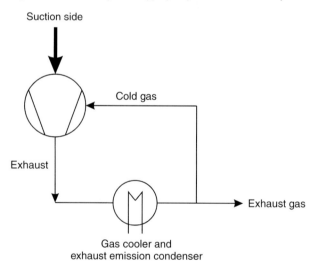

Figure 5.18 Cold gas circulation to cool down inner parts of this claw-type pump. (With permission from Dr.-Ing. K. Busch GmbH, Maulburg, Germany.)

The intake cycle begins by opening intake slot 2 due to the movement of the claw-type impellers. The process gas then flows into the steadily opening intake chamber of the first pumping stage. The intake of process gas functions, since a pressure gradient is produced through the increasing volume of this intake chamber. After precisely one turn of the impeller, the maximum pump chamber volume

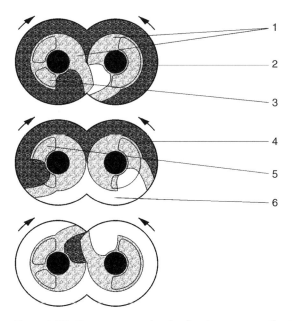

Figure 5.19 Pumping principle of a claw type pump without inner compression. (With permission from Dr.-Ing. K. Busch GmbH, Maulburg, Germany.)

is attained. As soon as the impeller has completed a turn, the intake slot closes and the intake process has been completed.

In parallel to each intake cycle, the gas which was in each case taken in before is compressed. This compression is effected by gas flowing backwards into the pump chamber (opening 3). As soon as one impeller opens the intake slot, the second impeller opens in the direction of gas inlet 3 on the exhaust side. The venting gas which is at atmospheric pressure flows into the pump chamber and increases the particle density in the pump chamber to the level of atmospheric pressure.

Since venting with cold gas is effected by a condenser related to the vacuum pump, intensive and effective cooling of the inner pump sections is attained. Finally the taken in gas and the warmed up cold gas is discharged through the exhaust gas slot 5 and is cooled down again in the downstream condenser (gas cooler) and freed of condensable substances.

This method of direct gas cooling was selected, because in connection with chemistry processes it is highly desirable to maintain temperatures which are not too high at the inside surfaces of the vacuum pump in contact with the gas, since the pumped gases and vapours, when exposed to high temperatures tend to thermally carbonize, thereby forming deposits, and also for safety engineering reasons the ignition temperatures of the usually combustible vapours which are pumped shall not be attained.

An important characteristic of this claw vacuum pump is, that the two shafts which drive the claw impellers, are exclusively suspended within the gear. A bearing on the pump chamber side does not exist. Through this simple design, this type of pump is particularly easy to service, all active pump sections are so designed that they can be pulled away towards the top.

5.3.5.3 New Developments of Screw Vacuum Pumps for the Area of Process Chemistry

The Screw Spindle Principle Newer developments of dry compressing process chemistry vacuum pumps are all based on the screw principle [3]. They employ as the displacers in each case a pair of screw spindles. Typical 'Roots impellers' have here been 'twisted' to resemble the shape of a screw. Here the passages form together with the casing wall sealed chambers which owing to the revolving spindle motion 'move' from the pump's inlet to the exhaust thereby conveying the gas. Losses due to backflows are limited by the fact that the gas on its way back would have pass through each chamber. The effect is similar to that of a labyrinth seal. This sealing effect is, moreover, augmented by differently high rotational speeds of the displacer of up to 8000 rpm which vary from manufacturer to manufacturer. When disregarding possible effects through high turbulences in the slots at higher rotational speeds, the ratio of pumped gas volume and backflow losses is naturally more favourable at higher rotational speeds, and thus the losses are lower.

An indicator for low backflow losses is the attainable ultimate vacuum with respect to the rotational speed. The ultimate vacuum attainable with these types of vacuum pumps amounts to 10^{-3} mbar (abs.).

Fundamentally, the screw spindles in modern screw vacuum pumps can be arranged as follows [4].

One advantage of the screw principle is immediately apparent: the gas is conveyed only in one direction without reversal or a change in direction. Any condensate which may have formed can be discharged relatively easily and continuously. Moreover, all screw vacuum pumps on the market, provided they are connected on their exhaust side to a permanently installed piping system, are relatively quiet and can thus also be installed in the vicinity of workplaces which are in constant use.

Screw vacuum pumps used in chemistry applications are being manufactured by different manufacturers (Figure 5.20).

A screw vacuum pump of the type a (Figure 5.21) known under the name of 'Cobra' is available on the market.

The Vacuum Pump is working with a pair of 'Archimedes type' screw rotors.

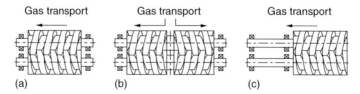

Figure 5.20 Different possibilities to arrange screw rotors in dry vacuum pumps. a, suspended on two sides; b, double flow and c, suspended on one side (cantilevered).

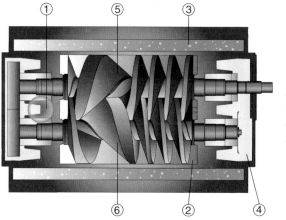

Figure 5.21 Sectional drawing of the Screw pump 'Cobra.' (With permission from Dr.-Ing.K. Busch GmbH, Maulburg, Germany.)

Main components:

1 = Inlet
2 = Exhaust
3 = Water cooling jacket
4 = Gear box, and
5 + 6 = Pump screw rotors.

Figure 5.22 Screw type pump 'Cobra' ready to be used in chemical processes. (With permission from Dr.-Ing. K. Busch GmbH, Maulburg, Germany.)

The contour of the screw impellers has been optimised for a high compression ratio. Thus an ultimate pressure down to 1×10^{-2} mbar is attained. The impellers are driven through the oil covered gear which is separated by shaft sealing rings from the pump chamber. The place of the shaft sealing rings may, if required and depending on the type of application, be taken up by four gear-oil-lubricated or even by entirely dry, gas-lubricated axial face seals (mechanical seals). As to the gear oil, the full range from mineral-oil-based to chemically entirely inert PFPE (perfluorinated polyether oil) can be used depending on the type of application (Figure 5.22).

The vacuum pump is cooled by means of a water jacket. In order to reduce the temperatures inside it is possible to admit cooling gas which can be adapted to the process conditions. For example, the cold gas can be taken downstream from a downstream condenser. But also fresh gas like nitrogen can be used for cooling. This dry compressing pump is not equipped with any valves. Any screw mechanism will be found to be quite insensitive to liquids entrained in the gas from the process. Liquid knocks can be handled quite well by the screw vacuum pump, but in connection with any kind of screw vacuum pump it needs to be ensured that no liquid can penetrate the gear chamber. This can be assured by use of mechanical seals as shaft seals. The dry screw pump Cobra is able to be controlled in vacuum by variation of the rotational speed (frequency controller). This possibility helps not only to have a good adjustment of the suction speed following the process needs, it helps as well to protect the pump against too high mechanical stress in case of sudden liquid knocks coming from the process (by automatic reduction of rotational speed via the frequency controller).

Another screw spindle pump of type 'c' (with cantilevered bearings) is available on the market under the name of Sihidry (Figure 5.23).

In the case of this screw spindle vacuum pump, the aspect of heat dissipation and limiting the temperature at the inside pump walls in contact with the gas was considered in great detail. Besides the usual jacket cooling arrangement, the cooling fluid also flows through the inner impellers. At the time this vacuum pump

Figure 5.23 Screw type vacuum pump 'Sihidry.' (With permission from Sterling SIHI GmbH, Itzehoe, Germany.) Shown is the service concept for easy cleaning of screw rotors. Main components: water cooled housing, two vertical screw rotors (internally cooled), driven by two electronic synchronized motors, gas flow top-down.

was introduced onto the market, this principle was entirely new in connection with dry compressing vacuum pumps.

This type of cooling arrangement is very helpful and highly welcome in all cases where the pumped gas or the vapour tends to thermal carburisation or polymerisation or exhibits a low ignition temperature. This type of cooling is also utilised in connection with air-cooled screw vacuum pumps new on the market [4].

The latest development in the area of screw spindle vacuum pumps (Figure 5.24).

Modern dry vacuum screw pumps are often used in processes, where polymerizing agents can reach the pump rotors. The polymers/oligomers over the operating time give rise to the growth of deposits. When using a modern screw vacuum pump, unscheduled downtimes owing to the excessive formation of deposits can, however, be reliably prevented. Modern screw vacuum pumps are equipped with a sensor monitoring system where the constant evaluation of casing vibrations, for example allows for the early detection of forming deposits. After a warning threshold has been exceeded, scheduled cleaning of the pump can be planned. The uninvolved cleaning work can then be performed on-site. This modern monitoring system also performs the detection of safe operating conditions by monitoring further parameters like temperatures and pressures within the scope of modern explosion protection measures in accordance with ATEX, by detecting inadmissible operating conditions before the occurrence of possible ignition sources, for example thereby allowing suitable measures to be introduced. Thus in connection with modern dry compressing screw vacuum pumps, generally the installation of flame arresters on the intake and the exhaust sides will no longer be necessary in all cases, as was the case for the first generation of dry compressing claw vacuum pumps, where the installation of such facilities was still mandatory in almost all cases. But the final recommendation must always be part of a good trained sales force of the vacuum pump manufacturers and this good sales force belongs to the complete package.

Reliable vacuum pump for chemical application and good knowledge about installation of these pumps, recommended accessories and after sales service.

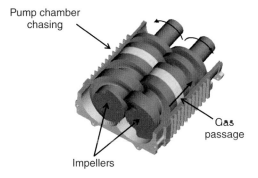

Figure 5.24 Principle of a horizontal screw vacuum pump for industrial applications type c (Figure 5.24). (With permission from Oerlikon Leybold Vacuum, Cologne, Germany.)

5.3.5.4 Outlook as to the Future of the Mechanical Vacuum Pumps in the Area of Chemical Process Engineering

From the pump manufacturers point of view, screw vacuum pumps are, as in the past, engineering-wise highly demanding machines which are constantly being developed further [5]. For the user, they can very much simplify and solve many application problems. For this reason they are becoming ever more popular and widespread in the area of chemical and pharmaceutical process engineering. With increasing numbers and advanced manufacturing techniques, the difference in price with respect to oil-sealed pumps and also to liquid ring vacuum pumps operating in closed cycles, will continue to drop. This will very much accelerate further spreading of dry compressing screw vacuum pumps. Future development targets in connection with these modern screw vacuum pumps for both industrial and chemistry process engineering applications can be described as follows:

- Reduction in the number of components in pumps with a high number of stages and thus more favourable manufacturing costs.
- Perfectioning of an on-site pump service which is as simple as possible.
- Fundamentally, the screw principle allows particles and formed condensate to be conveyed well through the pump, whereby this needs to be further optimised combined with ever improved purging capabilities for the process vacuum pumps.
- Low noise level so that the installation location can be selected freely.
- Low, but not too low inside temperatures of the surfaces in contact with the gas on the one hand so as to avoid thermal carburisation or even the ignition of pumped vapours and to avoid on the other hand early condensation within the pump. The avoidance of condensation is in all modern dry compressing pumps supported by measures already invented by one of the most important engineering personalities, namely Prof. Gaede from Cologne in the first decades of the past century, the gas ballast (see also Figure 5.8).

Owing to the ongoing dynamic further development as in the past which also needs to be assumed in the future for the area of dry compressing vacuum pumps for industrial process engineering, this chapter will have to be updated frequently also over the next years and decades.

References

1. Anonymous DIN 28400, Part 2.
2. Eckle, F.J., Jorisch, W., and Lachenmann, R. (1991) Vakuumtechnik im Chemielabor. *Vakuum in der Praxis*, **2**, 126–133.
3. Anonymous (1995) *Vacuum Technology for Chemical Engineering*, Brochure Published by, Oerlikon Leybold Vacuum.
4. Umrath, W. *et al.* (2007) *Fundamentals of Vacuum Technology*, Oerlikon Leybold Vacuum, Cologne.
5. Dreifert, T., Zöllig, U., and Stahlschmidt, O. (2005) Screwline in industriellen Anwendungen. *VIP-Vakuum in Forschung und Praxis*, **17** (2), 87–90.

6
Basics of the Explosion Protection and Safety-Technical Requirements on Vacuum Pumps for Manufacturers and Operating Companies

Hartmut Härtel

The title of this chapter refers to the requirements on vacuum pumps intended for use in potentially explosive atmospheres which result from the European Directives 94/9/EC [1] (ATEX 95 [1]) and 99/92/EC [2] (ATEX 137 [1]).

6.1
Introduction

At first, general explanations to the explosion protection are given. The basic properties and characteristics of an explosive atmosphere and the dependences between the particular safety characteristics are explained.

Then, the requirements on manufacturers and operating companies of explosion-protected equipment as specified in the Directives 94/9/EC [1] and 99/92/EC [2] are defined.

The term explosion-protected equipment used in the following covers plants, equipment and protective systems intended for use in potentially explosive areas. Explosion-protected vacuum pumps belong to the explosion-protected equipment as well.

The Directive 94/9/EC [1] of 23 March 1994 on the approximation of the laws of the member states concerning equipment and protective systems intended for use in potentially explosive atmospheres predefines

- requirements on the design and the construction of equipment and protective systems,
- criteria for an assignment of equipment to equipment groups and categories,
- methods to the proof of the conformity of manufactured equipment and protective systems with the requirements of the directive and
- methods of examination and production quality assurance.

---

1) The numbers 137 and 95 are numbers of paragraphs of the contract to the foundation of the European Community on which basis the directives were passed. The numbers were changed repeatedly: 118–137 and 100 over 100a–95.

Vacuum Technology in the Chemical Industry, First Edition. Edited by Wolfgang Jorisch.
© 2015 Wiley-VCH Verlag GmbH & Co. KGaA. Published 2015 by Wiley-VCH Verlag GmbH & Co. KGaA.

The directive is addressed to manufacturers of equipment, which is intended for use in potentially explosive areas. The member states of the European Community had to apply the provisions of the Directive as of 1 March 1996. The directive represents as of 1 July 2003 a valid European law which the member states have to observe for equipment and protective systems, as defined in the directive, to be placed on the market and put into service. The Directive was implemented, for example into German law by the '11. Verordnung zum Produktsicherheitsgesetz (11. ProduktSV)' (Product Safety Act) [3], other European states made equivalent laws.

The Directive 99/92/EC [2] of 16 December 1999 predefines minimum requirements for an improvement of the safety and health protection of workers who can be put at risk by explosive atmospheres.

This directive is addressed to operating companies of equipment which is intended for use in potentially explosive atmospheres. It represents a valid European law as of 1 July 2003. In the member states of the European Community it became legal force by respective legal provisions and administrative instructions (in Germany, for example by the 'Betriebssicherheitsverordnung (BetrSichV)' (Operational Safety Code) [4]).

6.2
Explosion Protection

6.2.1
General Basics

An *explosion*, in the following explanations, means an abrupt exothermic oxidation reaction between combustible (flammable) substances and atmospheric oxygen which is caused by an effective ignition source.

Therefore, an explosion can only arise if the three components

- combustible (flammable) substance,
- atmospheric oxygen and
- effective ignition source

concur physically and at the same time (Figure 6.1).

Combustible substances and the air in which the substances, under atmospheric conditions, are finely dispersed as gases, vapours, mists or dusts, form an explosive atmosphere. An atmosphere consisting of combustible substances or substance mixtures and air under atmospheric conditions is regarded as an explosive atmosphere, if, after an ignition, combustion processes propagate to the complete unburned mixture.

An exothermic oxidation reaction within the range of an explosive atmosphere is bound up with a sudden rise in temperature and/or pressure.

To avoid an explosion, a temporal and/or spatial coincidence of these three components has to be prevented.

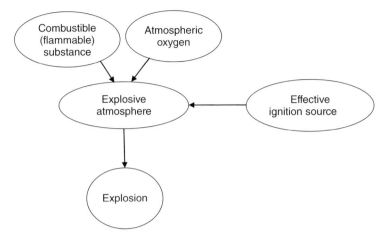

Figure 6.1 Conditions for the formation of an explosion.

Explosion protection means generally all measures for the avoidance of explosions and for the protection against explosion hazards.

The European directives for explosion protection, Directives 99/92/EC [2] and 94/9/EC [1], define measures for the protection against explosions of explosive atmospheres under atmospheric conditions. The two directives define an explosive atmosphere, as previously mentioned, as a mixture of air and combustible gases, vapours, mists or dusts under atmospheric conditions in which, after an ignition, combustion processes propagate to the complete unburned mixture.

However, the two directives do not define 'atmospheric conditions'.

Therefore, to define the atmospheric conditions the German regulations of the 'Berufsgenossenschaftlichen Explosionsschutz-Regeln BGR 104' (Explosion Protection Regulations of the Employer's Liability Insurance Association) [5] and the scope of application of the European Standard EN 13463-1:2001 [6] for non-electric apparatus are consulted.

The definition is: Atmospheric conditions exist if the absolute pressure-range is between 0.8 and 1.1 bar and the temperature is in the range from −20 to +60 °C.

The directives to the explosion protection do not cover

- explosions under conditions other than atmospheric conditions,
- explosions caused by decomposition of unstable compounds,
- explosions of explosives.

6.2.2
Explosive Atmosphere and Safety Characteristics

6.2.2.1 **General**
The properties of an explosive atmosphere are characterized essentially by the properties of the combustible substances or substance mixtures and by their

volume content in the explosive atmosphere formed by these substances or substance mixtures and air. Such combustible substances can be gases, vapours, mists or dust clouds, as already mentioned.

The further assessment refers essentially to explosive atmospheres consisting of air and combustible gases, vapours and/or mists. Explosive atmospheres, which consist of air and combustible dusts are normally not relevant for vacuum pumps and are not assessed here.

To form an explosive atmosphere the combustible substances must finely dispersed in air. Gases and vapours have mostly this ability. Mists, however, must form appropriately fine droplets so that a sufficiently great effective surface is maintained.

For a better assessment of the hazardousness of an explosive atmosphere, its properties are described by so-called safety characteristics. Knowledge of these safety characteristics allows a realization of appropriate measures for the avoidance of an explosive atmosphere.

Safety characteristics are not physicochemical material constants. However, they allow quantitative statements regarding the hazard potential that can result from an explosive atmosphere. They are the bases for the specification of measures for the fire and explosion protection.

The safety characteristics are determined in simple model experiments as practice-oriented as possible. The measured values depend on the determination methods, the technical parameters and the ambient conditions. Standardized measuring and test procedures are applied to achieve a comparability of the values.

The safety characteristics of a combustible substance/air mixture have to be determined in order to enable a better assessment of

- the explosibility,
- the ignitability caused by various ignition sources,
- the propagation of a reaction and
- the effects of an explosion.

The safety characteristics

- explosion limits and explosion range
- temperature and pressure limits for the stability of unstable substances
- explosion point and
- flash point

serve for an assessment of the explosibility.
The safety characteristics

- Minimum Ignition Energy (MIE) and minimum ignition current as well as
- Ignition temperature

are taken to assess the ignitability.
The propagation of a reaction is described by the following safety characteristics:

- limits of the detonation capability;
- propagation velocity of a deflagration and
- flame-proof gap width.

The effects of an explosion can be assessed by the following safety characteristics:

- maximum explosion pressure;
- maximum rate of pressure rise and
- effects of the pressure at detonations.

In the result of the determination of the safety characteristic 'Flame-proof gap width' and 'Ignition temperature' an explosive atmosphere can be assigned to an *Explosion Group* (Ex-Group) and a *Temperature Class* (T-Class).

For the practical application of safety characteristics, especially for the assessment of explosion hazards at pumping of explosive atmospheres with vacuum pumps, the different conditions for the determination of the safety characteristics and the operating conditions of the vacuum pumps have to be considered.

For example, a formal assignment of a T-Class on the basis of a surface temperature which is measured in the vacuum pump is not directly possible. The ignition temperature of a substance on which the T-Class classification is based is determined in a static medium under atmospheric pressure. However, the pumping media in the vacuum pump are only seldom under atmospheric pressure. Normally, they flow turbulently.

In the following, some safety characteristics are described further detailed.

6.2.2.2 Explosion Range and Explosion Limits

The *explosion range* is a concentration range of a combustible substance in air in which a mixture is explosive. That means that the mixture can be ignited by an ignition source at which the flame comes loose from the ignition source and propagates through the unburned mixture.

The *Lower Explosion Limit* and the *Upper Explosion Limit* (*LEL* and *UEL*) are concentration values of a combustible substance in air which define an *explosion range*.

Outside the explosion range, a combustible substance/air mixture will not explode either because of an oxygen deficiency or because of a deficiency of the combustible substance. An ignition will not propagate to the remaining unburned mixture. By removal of air oxygen, for example by a too-high content of combustible substance in the mixture or by inerting or a dilution of the mixture with air, a formation of an explosive atmosphere can be prevented.

In the context of an explosive atmosphere, a combustible substance/air mixture within the valid explosion limits is always meant.

Explosion limits are influenced by the pressure and temperature of an explosive atmosphere as well as by the intensity of an ignition source (ignition energy).

- An explosion range is reduced by a decrease of pressure of the combustible substance/air mixture. The LEL and the UEL will approach. An explosion range

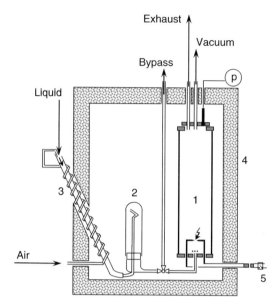

Annotation to Figure 6.2:
1. Explosion vessel with ignition electrode
2. Blending vessel
3. Vaporizer
4. Heating unit (drying oven)
5. Ignition source

Figure 6.2 Schematic representation of an apparatus for the determination of the explosion limits for explosive combustible substance/air mixtures [8].

disappears if the so-called *minimum ignition pressure* is reached. The LEL and the UEL merge. At a pressure below the minimum ignition pressure the mixture cannot be ignited any more.
- An explosion range is reduced by a decreasing temperature at a constant pressure. The minimum ignition pressure increases and vice versa.
- With an increasing ignition energy of the applied ignition source the explosion range is extended [7].

The maximum explosion pressure (P_{max}) which arises at an explosion and which is dependent on the initial pressure amounts to approximately the 8- to 10-fold of the initial pressure.

Figure 6.2 shows a schematic representation of an apparatus for the determination of explosion limits of combustible substance/air mixtures with different pressures as used by the PTB[2] for tests in a low pressure-range [8]. This apparatus allows tests at atmospheric pressure and pressures <1013 mbar and, within minor limits, variable temperatures. A standardized test method is described in [9].

2) PTB – Physikalisch-Technische Bundesanstalt Braunschweig and Berlin, Germany.

Figures 6.3 and 6.4 [8] show the dependence of the explosion range, including the minimum ignition pressure and the maximum explosion pressure, particularly at pressure less than or equal to the atmospheric pressure, on the mentioned influencing factors.

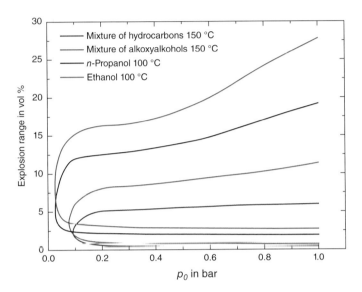

Figure 6.3 LEL and UEL for some solvents at approximately atmospheric pressure and pressures lower than atmospheric pressure [8].

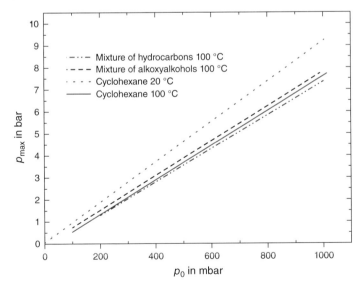

Figure 6.4 P_{max} depending on the initial pressure for some solvents at approximately atmospheric pressure and pressures lower than atmospheric pressure [8].

Figure 6.5 shows the vapour pressure (partial pressure) and the vapour concentration as a function of the liquid temperature of a flammable liquid [10]. The range D in Figure 6.5 shows the saturated vapour range. A saturated vapour exists only for a limited time. According to the vapour pressure, the vapour concentration descends to a concentration belonging to the liquid temperature. That is caused by condensation. In addition, the explosion limits, explosion points as well as the flash point are given in Figure 6.5. [7]. The LEL/UEL of a flammable liquid is the temperature related to the atmospheric pressure (1013 mbar) at which, under predefined test conditions, the concentration of the saturated vapour/air smixture above the liquid surface reaches the LEL/UEL. Therewith, the Lower Explosion Point (LEP) determines the lower limit temperature for the formation of explosive mixtures. If the vapour pressure curve and, for example the explosion limits of a pure liquid are known, the explosion points also can be assessed.

In a different way from the determination of the LEP, for the determination of the flash point in a 'closed cup' (see Section 6.2.2.3) the test vessel is opened for a short time for the immersion of the pilot flame. Air ingresses through the cover opening into the vapour chamber so that the vapour mixture is thinned slightly. In addition, the composition of the vapour/air mixture in the immediate environment of the pilot flame can be changed by arising combustion gases, for example.

The ignition tests for the determination of explosion points are carried out – similar to the determination of explosion limits – in a completely closed system with a stronger ignition source, a high-voltage spark gap.

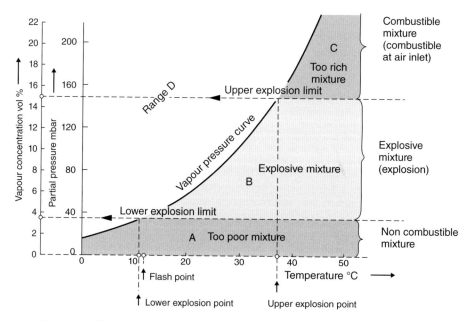

Figure 6.5 Vapour pressure and vapour concentration as a function of the liquid temperature for an ethanol vapour/air mixture, correlation between explosion limits and explosion points, representation of concentration ranges and their combustion properties [10].

Due to the test method, the flash points of pure substances are normally about 3–8 K higher than the LEP.

At atmospheres consisting of vapour of flammable liquids and air an ignition of the explosive atmosphere must already be anticipated at temperatures below the flash point, as of the LEP, in presence of an effective ignition source.

The determination of the LEP is then applied if, for example a determination of the flash point does not deliver any reproducible values or if it fails completely (vapourizing water or halogen contents of the mixture in the apparatus which influence the effectiveness of the ignition flame) or for the assessment of special safety-technical problems.

6.2.2.3 Flash Point

Besides combustible gases particularly the flammable liquids are very often involved in the formation of an explosive atmosphere.

The *Flash Point* is the lowest temperature of a flammable liquid at which – due to the respective vapour pressure – so much vapour arises above the liquid surface so that a flammable vapour/air mixture is formed.

Such a mixture could be ignited by a pilot ignition source. Flash points are determined according to standardized test methods in a closed or open cup. Usually, the flash point is determined in the closed cup according to Abel-Pensky, EN ISO 1523 [11] or according to Pensky-Martens, EN ISO 2719 [12]. The tests with an open cup (liquid surface is open to the environment) provide usually much higher flash points (up to 25 K) than the tests with a closed cup. This test method is mostly applied in the mineral oil industry. Figure 6.6 shows a schematic representation of

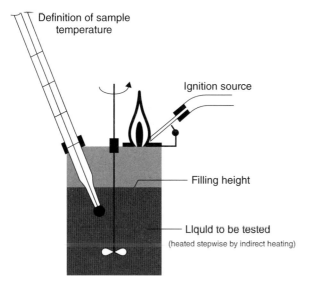

Figure 6.6 Schematic representation of an apparatus for the determination of a flash point with a closed cup according to Abel–Pensky [11].

a test apparatus for the determination of the flash point with a closed cup according to Abel–Pensky.

On the basis of the flash points determined in a closed cup and under consideration of the boiling point flammable liquids are classified according to their hazard qualities.

Mixtures of flammable liquids still may be classified according to the 'Gefahrstoffverordnung (GefStoffV)' (Hazardous Substances Regulation) [13] until May 2015 (see Table 6.1 [14]).

Criteria for a classification, labelling, and packaging (CLP) of flammable liquids according to the Regulation (EC) No. 1272/2008 [15] (CLP Regulation) can be found in Table 6.2. This classification is binding for substances as of 1 December 2010, and as of 1 June 2015, in addition, for mixtures/preparations. It corresponds to the criteria harmonized by the UN for the classification and labelling of substances and mixtures which have led to the Globally Harmonised System (GHS) of Classification and Labelling of Chemicals.

6.2.2.4 Minimum Ignition Energy and Ignition Temperature

The ignitability of a combustible substance/air mixture is determined by its MIE and the *ignition temperature* of the combustible substance. It must be

Table 6.1 Classification of flammable liquids according to the Hazardous Substances Regulation.

Flash point (°C)	Hazard property		
	Description of hazards	Risk symbol	R – phrase
<0 ($K_p \leq +35$)	Extremely flammable	F+	R 12
<+21	Highly flammable	F	R 11
+21 to +55	Flammable	—	R 10

Classification according to 'GefStoffV' [Hazardous Substances Regulation] [13] in combination with Directive 67/548/EEC [14]

Explanation: K_p = boiling point of the liquid.

Table 6.2 Classification of flammable liquids according to the CLP Regulation.

Classification according to Regulation (EC) No. 1272/2008 [15] (CLP Regulation)

Flash point (°C)	Initial boiling point (°C)	Category	Indication of hazardous properties
<23	≤35	1	H224: Liquid and vapour extremely flammable
<23	>35	2	H225: Liquid and vapour highly flammable
≥23, ≤ 60	—	3	H226: Liquid and vapour flammable

distinguished between the ignitability due to a *punctual* ignition source and a *laminar* ignition source.

Examples of punctual ignition sources are electrical sparks or mechanically generated friction or impact sparks. Examples of laminar ignition sources are hot machine parts or hot surfaces on equipment housings.

Minimum Ignition Energy By igniting with a punctual ignition source (spark ignition source) energy is supplied to a small mixture volume for a short time only. The quantity of energy must be at least so large that the small mixture volume ignites. However, a sufficient energy quantity must be available to warm the surrounding layers so strongly that the initial reaction can turn into a self-propagating reaction. The ignition energy required for such a reaction depends on the pressure, temperature and among others also on the concentration of the combustible substance/air mixture. The ignition energy limiting curve resulting from the concentration of the combustible substance at atmospheric pressure (1013 mbar) and approximately 20 °C shows a considerable minimum. This minimum value of the ignition energy is called Minimum Ignition Energy and the corresponding mixture concentration is described as 'the most readily ignitable mixture'.

Figure 6.7 shows the schematic representation of an apparatus for the determination of the MIE [16]. With this test procedure the lowest energy of a capacitive electrical circuit is determined whose discharge spark suffices to ignite the most readily ignitable combustible substance/air mixture (Table 6.3 shows the minimum ignition energies and most readily ignitable mixture concentrations of selected combustible gases and vapours.).

Ignition Temperature The ignition temperature is the lowest temperature of a hot surface at which the most readily ignitable combustible substance/air mixture.

Concerning an ignition by a laminar ignition source it is assumed that the combustible substance/air mixture has warmed itself in a larger area due to a longer retention time at a hot ignition source. At the beginning of the auto-ignition reaction the energy losses due to heat conduction in the surrounding layers and the hot area are already lower than the heat quantity produced in the reaction zone. The

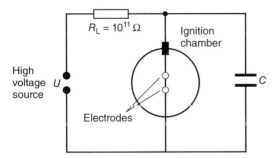

Figure 6.7 Schematic representation of an apparatus for the determination of the minimum ignition energy [16].

Table 6.3 Minimum ignition energy and most readily ignitable mixture concentration of selected combustible gases and vapours [16].

Substance	Minimum ignition energy MIE (mJ)	Most readily ignitable mixture (vol%)
Carbon disulfide (CS_2)	0.009	7.8
Hydrogen (H_2)	0.016	22.0
Acetylene (ethine – C_2H_2)	0.019	7.7
Ethylene (ethene – C_2H_4)	0.082	8.0
Methanol (CH_3OH)	0.140	14.7
Diethyl ether ($C_4H_{10}O$)	0.190	5.1
Benzene (C_6H_6)	0.200	4.7
Ethane (C_2H_6)	0.250	6.5
Propane (C_3H_8)	0.250	5.2
Methane (CH_4)	0.280	8.5
Acetone (C_3H_6O)	0.550	6.5
Ammonia (NH_3)	14	20.0
Trichloroethylene (C_2HCl_3)	510	26.0
Dichloromethane (CH_2Cl_2)	9300	18.0

oxidation reaction can therefore propagate independently. The temperature of a laminar ignition source required for a development of a self-propagating reaction is considerably lower than the required temperature of a point ignition source.

The characteristic value for the ignitability due to a laminar ignition source is the ignition temperature of a combustible substance/air mixture.

The ignition temperature depends on different factors, for example, the chemical composition and the concentration of combustible substance/air mixtures, the flow conditions, the material and the geometry of hot surfaces as well as the pressure at which the combustible substance/air mixture exists at a hot surface.

In practice, the respective conditions often cannot be described sufficiently exactly for a theoretical quantitative statement regarding an actually effective ignition temperature. Therefore, for the purpose of comparison, testing and standardization, the ignition temperature has been defined as a safety characteristic which can be determined reproducibly and which represents a minimum value. The apparatus for the determination of the ignition temperature according to EN 14522 [17] or DIN 51 794 [18] is schematically shown in Figure 6.8 [16]. This apparatus consists essentially of a 200-ml Erlenmeyer flask made of glass which is heated. The liquid to be tested is dosed drop by drop into the heated Erlenmeyer flask. For the determination of the ignition temperature of gases the gas to be tested flows into the test vessel (laminar flow). By variation of the temperature of the test vessel and the sample quantity the lowest temperature is determined at which an ignition with a just visible flame will occur.

Ignition temperatures which are usually determined at atmospheric pressure (1013 mbar) show a considerable pressure dependence. The ignition temperatures of the most substances decrease strongly in the pressure range between atmospheric pressure and about 2 bar (1 bar(gauge)) pressure. With further

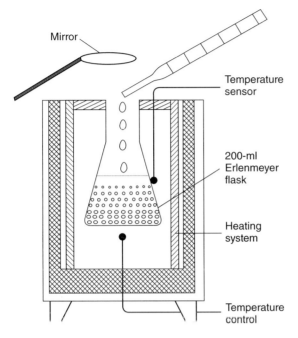

Figure 6.8 Schematic representation of an apparatus for the determination of the ignition temperature [16].

pressure increasing they approach to a value of approximately 200 °C. Figure 6.9 shows some typical curve progressions [19].

In the vacuum range, the ignition temperatures of the most flammable liquids rise strongly over the values, which are determined at atmospheric pressure.

The ignition temperatures determined with the standardized method are the basis for the classification of combustible gases and vapours into T-Classes pursuant to EN 60079-20-1 [20].

Table 6.4 contains the ignition temperatures of the respective T-Classes, the permissible surface temperatures for equipment of Category 2 and 3 and some examples of assigned substances.

6.2.2.5 Flame-Proof Gap Width

The ability of an explosive atmosphere to propagate the flames after an ignition also through narrow gaps is essentially described by the flame-proof gap width. The size of a flame-proof gap is particularly important for the assessment of the effectiveness of flame-proof encapsulations and for the resistance of flame arresters (FAs) against flame transmission. An apparatus for the determination of the maximum gap according to EN 60079-20-1 [20] is used for the determination of the flame-proof gap width of a combustible substance/air mixture. This apparatus contains two half-spheres in a cylindrical outer vessel. The half-spheres are separated by a 25-mm long adjustable joint gap. A high-voltage spark gap

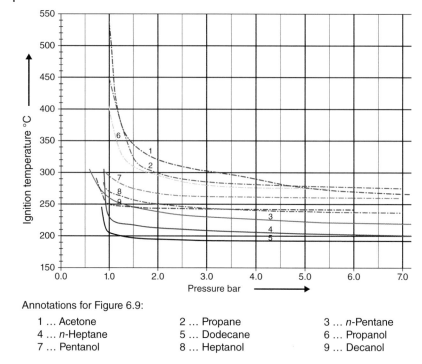

Annotations for Figure 6.9:
1 ... Acetone   2 ... Propane   3 ... n-Pentane
4 ... n-Heptane   5 ... Dodecane   6 ... Propanol
7 ... Pentanol   8 ... Heptanol   9 ... Decanol

Figure 6.9 Typical courses of curves of the ignition temperature as a function of pressure [19].

serves as electrical ignition source. It is placed inside the inner centre of the spherical explosion chamber.

The inner explosion chamber and the outer vessel are filled with the explosive mixture to be tested. By igniting the mixture inside the spherical chamber is tested whether an ignition is transferred through the joint gap to the outer vessel.

The maximum gap of a combustible substance/air mixture is the gap width at which an ignition through the gap does not occur just any more. The lowest value of all maximum gaps of a substance that was determined under atmospheric pressure in dependence on the substance concentration is the *Maximum Experimental Safe Gap* (MESG). The appertaining combustible substance mixture is designated as the '*most readily ignitable mixture*'. The gap width depends on the temperature and the pressure. A rise in pressure and temperature leads to a reduction of the gap width.

The MESG's determined for the individual substances serve exclusively for the classification of substances regarding their ability for transmission of an internal ignition. They are not a measure for the constructive dimensions of flame-proof gaps at the type of protection 'flame-proof enclosure' or for gap widths of FAs.

Figure 6.10 shows the dependence of the gap widths on the mixture concentration for different substances. A more or less distinctive minimum is recognized at

Table 6.4 Temperature classes pursuant to EN 60079-20-1 [20] and surface temperatures permitted for equipment according to EN 13463-1 [6] and examples of substances.

T-Class	Ignition temperatures (°C)	Permissible surface temperatures for equipment of Category 2 and 3 (°C)	Examples for combustible substances with ignition temperature (in brackets)
T1	>450–	450	Methane (595 °C), propane (470 °C) and hydrogen (560 °C)
T2	>300–450	300	n-Butane (365 °C) and n-propanol (405 °C)
T3	>200–300	200	Motor gasoline, diesel fuel, n-heptane (215 °C) and isoprene (220 °C)
T4	>135–200	135	Diethyl ether (170 °C) and ethanol (140 °C)
T5	>100–135	100	—
T6	>85–100	85	Carbon disulfide (95 °C)

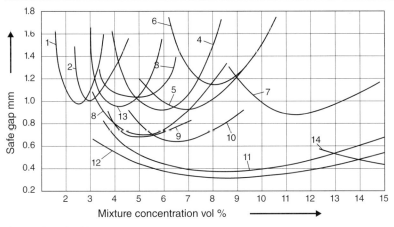

Annotations for Figure 6.10:
1 … n-Hexane
2 … Methylisobutylketone
3 … Ethyl acetate
4 … Ethane
5 … Acetaldehyde
6 … Methane
7 … Hydrogen sulfide
8 … Propylene oxide
9 … Dioxane
10 … Ethylene
11 … Acetylene
12 … Carbon disulfide
13 … Propane
14 … Hydrogen

Figure 6.10 Dependence of the safe gap on the mixture concentration for different substances [10]

all represented curves. This minimum indicates the MESG and the appertaining most readily ignitable mixture concentration of the respective substance.

A schematic representation of an apparatus for the determination of the MESG's of explosive combustible substance/air mixtures is shown in Figure 6.11. The PTB has used this apparatus for tests in the low pressure-range [8]. It can be used for

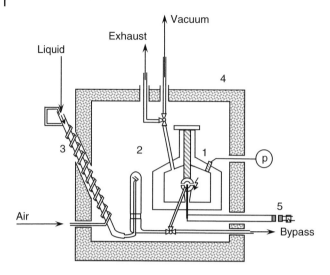

Annotations for Figure 6.11:
1. Explosion vessel with ignition source
2. Blending vessel
3. Vaporizer
4. Heating unit
5. Ignition device

Figure 6.11 Schematic representation of an apparatus for the determination of the MESG of explosive combustible substance/air mixtures [8].

tests at atmospheric pressure and pressures <1013 mbar and within a minor limit with variable temperatures as well.

Based on the MESG's determined at atmospheric pressure and according to the specifications in EN 60079-20-1 [20], substances are assigned to the Ex-Groups I, IIA, IIB and IIC. The standard EN ISO 16852 [21] for the constructive design and tests of FAs specifies the Ex-Group I as IIA1. The Ex-Group IIB is subdivided into four subgroups.

Table 6.5 gives an overview about the classification of the combustible substance/air mixtures into Ex-Groups according to their MESG's as per EN 60079-20-1 [20] and EN ISO 16852 [21].

6.2.3
Measures of the Explosion Protection

In accordance with the regulations of the Directives 99/92/EC [2] and 94/9/EC [1] the fundamental principle of the explosion protection must be implemented in order to maintain the explosion protection, that is the protection against hazards by explosions of explosive atmospheres. This fundamental principle covers three essential groups of measures of the explosion protection:

Table 6.5 Classification of combustible substance/air mixtures into Ex-Groups according to their MESG's as per EN 60079-20-1 [20] and EN ISO 16852 [21].

Classification as per EN 60079-20-1 [20]		Classification as per EN ISO 16852 [21]		Examples with MESG values (mm)
Ex-Group	MESG (mm)	Ex-Group	MESG (mm)	
I[a]	1.14	IIA1	≥1.14	Methane (1.14)
IIA	> 0.90	IIA	>0.90	Toluene (1.06)
IIB	0.50–0.90	IIB1	0.85–0.90	Diethyl ether (0.87)
		IIB2	0.75 to < 0.85	Ethylene glycol (0.78)
		IIB3	0.65 to < 0.75	Ethylene (0.65)
		IIB	0.50 to < 0.65	Ethylene oxide (0.59)
IIC	< 0.50	IIC	<0.50	Hydrogen (0.29)

a) Methane, for gassy mines.

1) measures, which prevent or restrict the formation of hazardous, explosive atmosphere (*primary explosion protection*);
2) measures, which prevent or restrict effective ignition sources (*secondary explosion protection*);
3) If explosions cannot be excluded for sure, technical measures which limit the effects of an explosion to a harmless extent (no personal injuries, tolerable damages to property) (*tertiary* or *constructive explosion protection*).

The mentioned measures altogether must reduce an *explosion risk* and keep it on an acceptable low level.

The following question arises: What does the term risk mean?

The *risk* (R) of a technical process or situation, which can occur in combination with an unwanted event, represents a factor that allows only qualitative information about the probability of an unwanted event (H) and *the probability of damage effects* (S) [10].

Nabert et al. [10] specifies the formula:

$$R = H \times S$$

The probability of an unwanted event and the probability of unwanted damage effects can only be multiplicatively combined if the two probability factors are independent of each other.

The *highest acceptable risk* (R_V) is a risk which is just acceptable. Usually, it can be specified only qualitatively. This is a discretionary decision and influenced by the respective plant and process conditions.

The *safety* of a technical process or situation is given if

$$R \leq R_V$$

is assumed.

There is no *absolute safety* of a technical process or situation ($R = 0$).

The probability of an unwanted event (H) of an explosion results from the product of the two place- and time-dependent factors *mixture probability* (H_G) and *probability of an ignition* (H_Z), which are independent of each other.

The mixture probability, that is the probability of occurrence of an explosive atmosphere, depends on type, quantity, composition and consistency of the combustible substances, their safety characteristics and possible release and propagation in the surrounding air.

The probability of an ignition, that is the probability that potential ignition sources become effective, depends on type, quality and mode of functioning of the operating supplies and work equipment.

From the fundamental principle of the explosion protection results the *basic concept of the explosion protection*.

The formula

$$R = H_G \times H_Z \times S \leq R_V$$

is valid [10].

As mentioned above, the probability of an unwanted event (H) of an explosion can be described by the mixture probability (H_G) and the probability of an ignition (H_Z).

H_G can essentially be influenced by measures of the primary explosion protection and H_Z by measures of the secondary explosin protection. Measures of the tertiary explosion protection affect principally the probability of unwanted damage effects (S).

Following these basic statements, the Directive 99/92/EC [2] describes the measures of the primary and tertiary explosion protection as a matter of priority. These measures are essentially aimed at operating companies/users of work equipment as well as workplaces in which potentially explosive atmospheres can occur.

The Directive 94/9/EC [1] aims mainly at constructors and manufacturers of work equipment intended for use in potentially explosive atmospheres. The secondary measures of explosion protection are discussed as a matter of priority in order to prevent effective ignition sources at the work equipment.

6.3
Directive 99/92/EC

6.3.1
Requirements on Operating Companies of Vacuum Pumps

The essential requirements on operating companies of explosion-protected equipment, for example of explosion-protected vacuum pumps, arise from the specifications of Directive 99/92/EC [2]. This directive fixes minimum requirements for improving the safety and health protection of workers who are potentially endangered by explosive atmospheres (section 1, article 1 of the directive).

The *obligations* for the *employers* are specified in section II of the directive:

- Prevention of explosions and protection against explosions (article 3 of the directive)
  In accordance with the basic principle of the explosion protection the employer must take technical and/or organisational measures in the following order of priority:
  - the prevention of the formation of explosive atmospheres, or where the nature of the activity does not allow that,
  - the prevention of the ignition of explosive atmospheres, and if the nature of activity does not allow that either,
  - the mitigation of a possible explosion, that is attenuation of detrimental effects of an explosion to ensure the safety and health of the workers.
- *Assessment of explosion risks* (article 4 of the directive) The employer shall assess the specific risks arising from explosive atmospheres in their entirety, taking account at least of
  - the likelihood that explosive atmospheres will occur and their persistence,
  - the likelihood that ignition sources, including electrostatic discharges, will be present and become active and effective,
  the installations, substances used, processes and their possible interactions and
  - the scale of the anticipated effects of an explosion.
- *General obligations of the employer* to ensure the safety and health of workers (article 5 of the directive)Considering the basic principles of risk assessment the employer must take the following measures:
  - The working environment where explosive atmospheres may arise has to be organized so that work can be performed safely.
  - During the presence of workers in working environments where explosive atmospheres may arise, the supervision of the working environments has to be ensured in accordance with the risk assessment by the use of appropriate technical means.
- Classification of places where explosive atmospheres may occur (article 7 of the directive)
  Places where explosive atmospheres may occur:
  - The employer must classify these places into zones in accordance with annex I of the directive.
  - The employer must ensure that the minimum requirements laid down in annex II of the directive are applied to these places.
  - The points of entry to places where explosive atmospheres may occur and which have been assigned to zones in accordance with annex I of the directive, must be marked according to annex III of the directive.
- Drawing up an explosion protection document (article 8 of the directive)In the scope of the obligation to the assessment of explosion risks (article 8 of the directive) the employer must draw up an explosion protection document and keep it up to date.

The explosion protection document must demonstrate in particular
- that the explosion risks have been determined and assessed,
- that adequate measures are taken to attain the aim of the directive – the explosion protection,
- those places which have been classified into zones in accordance with annex 1, and that in these places the minimum requirements set out in annex II are fulfilled,
- that workplaces and work equipment, including warning devices, are designed, operated and maintained with respect to safety,
- that in accordance with Directive 2009/104/EC [22] (superseding Directive 89/655/EWG) all necessary precautions have been made for the safe use of work equipment.

The explosion protection document must be drawn up *prior* to the commencement of work!

- Observance of special regulations for work equipment and workplaces (article 9 of the directive)
  - Work equipment for explosive atmospheres which is already in use or is made available for the first time before 30 June 2003 must comply as of 1 July 2003 with the minimum requirements specified in annex II, chapter A of the directive.
  - Work equipment used in potentially explosive atmospheres which is used for the first time after 30 June 2003 must comply with the minimum requirements specified in annex II, chapters A and B of the directive. That means that they must comply with the requirements of Directive 94/9/EC [1] among others.
  - Workplaces where explosive atmospheres may occur and which are used for the first time after 30 June 2003 must comply with the minimum requirements set out in the directive.
  - Workplaces where explosive atmospheres may occur and which are already in use before 30 June 2003 must comply with the minimum requirements laid down in the directive not later than three years after that date.
  - After 30 June 2003, any modification, extension or restructuring must be undertaken so that they comply with the minimum requirements of the directive.

The minimum requirements of Directive 99/92/EC [2] must be fulfilled in all member states of the European Community. If necessary, further requirements can be specified according to the national conditions.

In summary the following measures to maintain the explosion protection result for the operating company of workplaces and work equipment which can be endangered by explosive atmospheres as the *most essentially obligations*:

- Implementation of the basic concept of the explosion protection
  - Prevention of a formation of explosive atmospheres,
  - Avoidance of an ignition of explosive atmospheres,
  - Mitigation of explosion effects.

- Determination and assessment of remaining explosion risks,
- Safe organisation and supervision of the working environments where explosive atmospheres may arise,
- Classification of hazardous areas into zones, observance of the corresponding minimum requirements for the respective zone and marking of potentially explosive areas,
- Drawing up an explosion protection document,
- Observance of scheduled dates for the organisation of workplaces and work equipment.

6.3.2
Classification of Hazardous Areas into Zones

Hazardous areas are classified into zones according to the specifications in annex I of Directive 99/92/EC [2].

The classification is based on the type of explosive atmosphere. Atmospheres are formed from

- air and combustible *gases*, *vapours* and *mists* (zones 0, 1, 2) and
- air and combustible *dusts* (zones 20, 21, 22).

Hazardous areas are areas where explosive atmospheres may arise in such quantities so that special protective measures for the maintenance of the safety and health protection of workers have to be taken.

Areas where explosive atmospheres in such quantities are not expected are designated as 'non-hazardous areas'.

Hazardous areas can be found in workplaces, that is in the environment of work equipment, and also inside of plants, equipment and devices.

Referring to the explanations to explosion risks in Section 6.2.3, the following system of classification of zones represents a qualitative assessment of the mixture probability H_G according to duration and frequency of an occurrence of an explosive atmosphere. A real quantitative assessment of the mixture probability is only hardly possible in practice.

The *zones 0, 1 and 2* as well as *20, 21 and 22* stand for

- a very *high*
- a *middle*
- a very *low*

probability of an occurrence of explosive atmospheres.

The criteria for the determination of zones are as follows:

Zones with explosive atmospheres which are formed from *air* and combustible *gases, vapours* or *mists*:

Zone 0: Explosive atmosphere is present *continuously, for a long periodor frequently.*	Zone 1: Explosive atmosphere *is likely to occur in normal operation occasionally.*	Zone 2: Explosive atmosphere *is not likely to occur in normal operation,* but if it does occur *for a short period only.*
Zone 20: Explosive atmosphere in the form of a *cloud* of combustible dust in air *is present continuously, for a long period or frequently.*	Zone 21: Explosive atmosphere in the form of a *cloud* of combustible dust in air *is likely to occur in normal operation occasionally.*	Zone 22: Explosive atmosphere in the form of a *cloud* of combustible dust in air *is not likely to occur in normal operation,* but if it does occur *for a short period only.*

Annotations:

1) Layers, deposits and accumulations of combustible dusts must be considered like any other source, which can form an explosive atmosphere (dispersed dust forms clouds).
2) 'Normal operation' means the situation when installations and plants are operated within their design parameters.

6.4
Directive 94/9/EC

6.4.1
Equipment Groups and Categories

The Directive 94/9/EC [1] includes essentially the requirements for a safe design and construction of equipment and protective systems intended for use in potentially explosive atmospheres.

In accordance with chapter I, article 1, the directive applies both to electrical and non-electrical equipment and protective systems.

Besides equipment and protective systems it applies, in accordance with article 1, subitem 2, to safety devices, monitoring devices and controlling systems intended for use outside potentially explosive atmospheres but required to ensure the explosion protection of equipment and protective systems.

In accordance with the directive, equipment is assigned to various *equipment groups* and *categories*:

- *Equipment group I*: Equipment intended for use in underground parts of mines and surface installations of mines that may be endangered by firedamp and/or combustible dusts.
- *Equipment group II*: Equipment intended for use in other areas that are may be endangered by explosive atmospheres.

Referring to the explanations to the risk in Section 6.2.3, the classification of equipment into categories means a qualitative assignment according to the ignition probability H_Z which arises from potential ignition sources inside and/or outside the equipment.

The *equipment group I* is subdivided into the categories M1 and M2 with a very low ignition probability and an increased ignition probability.

- *Category M1*: Equipment is required to remain functional in the event of an explosive atmosphere.
- *Category M2*: Equipment is intended to be de-energized in the event of an explosive atmosphere.

The *equipment group II* comprises the categories 1, 2 and 3 for equipment with increasing ignition probabilities described as follows:

- *Category 1*: In the event of *rare* failures *no* ignition source will become effective.
- *Category 2*: Effective ignition sources are likely to occur in the event of *rare* disturbances or equipment failures.
- *Category 3*: Effective ignition sources are likely to occur at *anticipated* disturbances or equipment failures, but not during normal operation.

Exact criteria for the classification into categories can be found in the annex I of Directive 94/9/EC [1].

The following overview represents the essential properties and requirements for equipment of group II for their assignment to categories.

The requirements for explosion protection on equipment of group I as well as on equipment of group II for mixtures of combustible dusts and air are not further described here.

6.4.1.1 Equipment Group II

Category 1:

The equipment is capable to ensure *a very high level of protection in the event of rare disturbances or equipment failures.*

- In the event of failure of one technical measure of protection, at least a second technical measure of protection must provide the requisite level of protection.
- In the event of two faults occurring independently of each other the requisite level of protection must be assured.
- Ignition sources must be avoided even at rare equipment failures.
- The specified highest surface temperatures must not be exceeded in the worst case either.

Category 2:

This equipment is capable to ensure *a high level of protection* even *in the event of anticipated, frequently occurring disturbances or equipment failures.*

- Ignition sources must be avoided even at frequently occurring disturbances or equipment failures.
- The specified highest surface temperatures must not be exceeded in the event of unusual operating situations assumed by the manufacturer.

Category 3:

This equipment is capable to ensure *a normal level of protection during normal operation.*

- Foreseeable ignition sources which can occur during normal operation have to be avoided.
- The specified highest surface temperatures must not be exceeded during an intended operation. An overrun of limit temperatures is only permitted if the manufacturer has provided additional special measures of protection.

6.4.2
Assignment between Equipment Categories and Zones

To hold the risk (R) of an event, that is explosion risk, with an acceptable economic effort below the specified limit risk R_V ($R \leq R_V$), the probability of an unwanted event (H) of such event must be low at a predefined probability of damage effects (S).

Because H results from the product of the values H_G and H_Z, the ignition probability H_Z must be low at a high mixture probability H_G.

The following results:

- Equipment with a very low ignition probability (equipment category 1) must be used in areas with a high mixture probability (zones 0 and 20).
- Equipment with a higher ignition probability (equipment category 3) can also be used in areas with a lower mixture probability (zones 2 and 22).

The *assignment* between *zones* of potentially explosive areas and the applicable *equipment categories* results from this knowledge (see Table 6.6).

Table 6.6 Overview about the assignment between equipment categories and zones for the implementation of the explosion protection for the equipment group II.

Equipment category		Applicable for zones	Explosive atmosphere formed from air and combustible substances as
1	1G	0, 1, 2	Gases, vapours, mists
	1D	20, 21, 22	Dusts
2	2G	1, 2	Gases, vapours, mists
	2D	21, 22	Dusts
3	3G	2	Gases, vapours, mists
	3D	22	Dusts

6.4.3
Requirements on Manufacturers of Vacuum Pumps

The requirements on manufacturers of explosion-protected vacuum pumps result from the regulations of Directive 94/9/EC [1] for equipment and protective systems intended for use in potentially explosive areas.

In accordance with article 3 of the directive, all equipment and protective systems as well as devices mentioned in the article 1, section 2 of the directive must meet the *essential safety and health requirements* set out in the annex II of the directive.

The *essential requirements* comprise

- 'common requirements for equipment and protective systems',
- 'supplementary requirements on equipment', depending on category and
- 'supplementary requirements on protective systems'.

The basic requirements of the directive are concretised by harmonised European standards. A list of the valid harmonised standards is regularly updated and published by the European Commission in the official journal of the EU. The relevant standards for non-electrical equipment (e.g. vacuum pumps) which define the requirements of the directive are, for example the EN 1127-1 [23], the series EN 13 463 [6–24] and EN ISO 16852 [21].

Among other things, these *standards* demand a *risk assessment* and *ignition hazard analysis* for each equipment or protective system in the meaning of the directive.

With the results of a risk assessment and an ignition hazard analysis the manufacturer of equipment intended for use in potentially explosive atmospheres must assign the equipment to the respective equipment category.

A *compliance* of the manufactured equipment and protective systems with the requirements of the harmonised European standards allows the *assumption* that the essential safety and health requirements of Directive 94/9/EC [1] are fulfilled.

6.4.4
Conformity Assessment Procedure

The *conformity assessment procedure* for the proof of the compliance of explosion-protected equipment and protective systems to be supplied with the regulations of the directive is described in annex II, article 8 of the directive.

Depending on equipment group, category and design of equipment and protective systems, the conformity assessment procedure defines

- the tests to which equipment and protective systems must be subjected and
- the necessary measures for a quality assurance in the production

to prove the conformity with the directive before the manufacturer can issue an EC-Declaration of Conformity for the supply of equipment and protective systems.

In accordance with the conformity assessment procedure the *protective systems* have to be subjected to an EC-type examination according to annex III of the directive by a Notified Body.

Notified Bodies are national testing authorities, which have been notified by the respective member state of the European Community for the execution of the proceedings in the context of the conformity assessment procedure. The Notified Bodies and their identification numbers have to be informed to the European Commission.

Vacuum pumps on which various electrical and non-electrical equipment or accessories can be installed are regarded as *non-electrical equipment*.

The pump outside is normally assigned to category 2 or 3 because the installation site of the vacuum pump is classified as zone 1 or 2, if the risk assessment according to Directive 99/92/EC [2] fixes nothing else for this workplace.

The pump inside is assigned to the categories 1, 2 or 3. This depends on the zone (zone 0, 1 or 2) from which the atmosphere is pumped.

In accordance with the conformity assessment procedure according to article 8 of the directive, the *vacuum pumps of category 1* that pump from zone 0 have generally to be subjected to an EC-type examination. If the examination has been passed with a positive result the Notified Body issues an EC-type examination certificate.

The quality assurance can be carried out with the procedures as described in the annexes IV or V of the directive.

Vacuum pumps of category 2 which pump from zone 1 have to be subjected by the manufacturer to an internal control of production according to annex VIII of the directive. In addition, the manufacturer has to deposit a technical documentation for the vacuum pump at a Notified Body, including a risk assessment, ignition hazard assessment and notes for the implemented measures to maintain the explosion protection.

Vacuum pumps of category 3, which pump from zone 2 and which are operated in zone 2, only have to be subjected to an internal control of production according to annex VIII of the directive by the manufacturer.

Independently of the category, equipment and protective systems can be subjected to *unit verifications* according to annex IX of the directive by a Notified Body. If the unit verification has gone with a positive result the Notified Body issues a *declaration of conformity*. This procedure is applied if only few equipments or series with a little number of items are produced.

6.4.5
Application of the Regulations of the Directive

Resulting from the previous *practical implementation of the regulations* of the Directive 94/9/EC [1] the following can be stated for vacuum pumps:

- Vacuum pumps which are intended for operation in potentially explosive atmospheres have to be subjected to a risk assessment and ignition hazard analysis according to the procedure described in the Section 6.4.4.
- Depending on the results of the risk assessment and ignition hazard analysis the measures for explosion protection for the wanted equipment category must be realized.
- A vacuum pump must fulfil at first the requirements of standard EN 1012-2 [25] regarding resistance to pressure and tightness.
- The standard EN 1012-2 [25] defines in general the regulations of the EC-Machinery Directive 2006/42/EC [26] for vacuum pumps.
- To avoid effective ignition sources inside the vacuum pumps and at the outer parts of the vacuum pumps, the requirements of the standard series EN 13 463 [6–24] have to be fulfilled and the mentioned types of protection have to be applied.
- Because in spite of the application of the above-mentioned types of protection the effective development of potential ignition sources at positive displacement pumps cannot be excluded completely in the event of rare operating disturbances and equipment failures, the remaining risk must be minimized by application of explosion decoupling measures according to EN 1127-1 [23]. Explosion decoupling means here especially the installation of FAs at the pump inlet and pump outlet and the explosion-resistant design of the vacuum pump.
- Pursuant to the usual practice the positive displacement pumps of the category 1 which pump from the zone 0 must be equipped with FA's at the inlet and outlet side and they must be in an explosion-resistant design.
- The proof of the resistance to flame transmission of the FA, the proof of the explosion-resistant design and the specification of the permissible T-Classes inside and outside the pump is delivered within an EC-type examination by a Notified Body.
- Whether vacuum pumps of the category 2 which pump from the zone 1 have to be equipped with FA's and whether they must be in an explosion-resistant design is dependent on the results of the risk assessment and ignition hazard analysis of the manufacturer.

- If it can be proved beyond all doubt that effective ignition sources are expected only at rare operation disturbances and equipment failures, an explosion-resistant design and an installation of FA's is not required for these vacuum pumps.
- The manufacturer of the vacuum pump comes to a decision for the renunciation of FA's and an explosion-resistant design.
- If, due to the result of the risk assessment and the ignition hazard analysis, vacuum pumps of the category 2 must be equipped with FA's and if they must be in an explosion-resistant design, then the experimental proof of the resistance to flame transmission of the FA's can be delivered together with the vacuum pump, like for the category 1.
- After the conclusion of the conformity assessment procedure and of the tests perhaps required, the manufacturer issues an EC-declaration of conformity for his product. He affixes the explosion protection marking (Ex marking) to the product and if the product fulfills the requirements of other applicable European directives he affixes a CE marking to the product as well.
- The Ex marking for protective systems and equipment of the category 1 is normally determined by the Notified Body according to the result of an EC-type examination. The Ex marking for non-electrical equipment of the categories 2 or 3, for example explosion-protected vacuum pumps, is determined by the manufacturer or by the supplier of this equipment.

Examples for Ex marking:

- Protective system, for example FA as an in-line FA, for explosive atmosphere formed from air and gases, vapours and/or mists of the Ex-Group: IIA, maximum temperature of the FA: +60 °C, ambient temperature (T_a): $-20\,°C \leq T_a \leq +40\,°C$ (standard values):

- Non-electrical equipment of the equipment group II, category 2 for explosive atmosphere formed from air and gases, vapours and/or mists, type of protection: flame-proof enclosure, permissible explosive atmosphere of the Ex-Group: IIB, T-Class: T4, ambient temperature (T_a) with: $0\,°C \leq T_a \leq +60\,°C$ differing from the 'standard values': $-20\,°C \leq T_a \leq +40\,°C$:

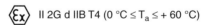

- Vacuum pump of equipment group II, category 1 inside, category 2 outside, for explosive atmosphere formed from air and gases, vapours and mists, type of protection: protection by constructional safety, in the inside: permissible explosive atmosphere of the Ex-group: IIB3, T-Class: T3, outside: Ex-Group: IIB, T-Class: T4, ambient temperature (T_a): $-20\,°C \leq T_a \leq +40\,°C$ (standard values):

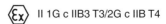

A *CE marking* is affixed to the product in addition to the *Ex marking*. The marking of protective systems and non-electrical equipment of the category 1 it is completed with the *Identification Number of the Notified Body*, who has certified the quality assurance system in accordance with the annexes IV or V of the Directive 94/9/EC [1].

6.5 Summary

Explosive atmosphere under atmospheric conditions is formed from air and combustible substances. The combustible substances are finely dispersed in the air as gases, vapours, mists or dusts.

The properties of an explosive atmosphere are determined essentially by the properties of the combustible substances and their volume content in the explosive atmosphere as well as by the pressure and temperature of the explosive atmosphere.

Safety characteristics are determined for the characterization of an explosive atmosphere. They are ascertained with standardized test methods. They are not physicochemical substance constants, because the values distinctly depend on the respective test conditions. Explosive atmospheres are classified into T-Classes on the basis of the safety characteristic 'Ignition Temperature'. They are classified into Ex-Groups on the basis of the safety characteristic 'MESG'. When the safety characteristics are applied in practice, the differences between the test conditions and the practical conditions have always to be taken into account.

The European Directives 99/92/EC [2] and 94/9/EC [1] contain the basic requirements for the prevention of explosions which can be caused by explosive atmospheres. The Directive 99/92/EC [2] points as a matter of priority to the operating companies of plants and equipment for a safe design of workplaces and processes. The Directive 94/9/EC [1] defines the basic requirements for a safe design of equipment and protective systems (i.e. free of ignition sources and/or poor in ignition sources) which are used in potentially explosive atmosphere.

Before vacuum pumps are put into operation, the operating company of a vacuum pump has to examine and assess the explosion risks which can be caused in the connection with the installation site and the processes applied. He must prevent or limit the formation of explosive areas. Remaining explosive areas have to be assigned to zones. Possible ignition sources in the zones must be avoided. The work equipment for the required equipment categories, for example explosion-protected vacuum pumps of the respective category, has to be selected. The results of the risk assessment, the applied protective measures, the zone classification and the remaining risks have to be recorded in the explosion protection document.

The manufacturer of work equipment intended for use in potentially explosive atmospheres has to construct and manufacture his products in compliance with the basic requirements of the Directive 94/9/EC [1]. The products have to be subjected to a risk assessment and ignition hazard analysis. In the result of them the products have to be assigned to the corresponding equipment category. In accordance with this equipment category the manufacturer has to introduce the defined conformity assessment procedure.

If the conformity of the product with the specifications of the Directive 94/9/EC [1] or other applicable directives, for example Machinery Directive 2006/42/EC [26] and EMV-Directive 89/336/EWG [27], is proved, the manufacturer has to provide his product with the respective Ex Marking and CE marking before the product will be supplied.

References

1. Directive 94/9/EC of the European Parliament and the Council of 23 March 1994 on the approximation of the laws of the Member States concerning equipment and protective systems intended for use in potentially explosive atmospheres, last corrigendum: 26 January 2000. *Off. J. EC*, **21**, 42.
2. Directive 99/92/EC of the European Parliament and the Council of 16 December 1999 on minimum requirements for improving the safety and health protection of workers potentially at risk from explosive atmospheres, last corrigendum of 7 June 2000. *Off. J. EC*, **134**, 36.
3. Elfte Verordnung zum Produktsicherheitsgesetz (Explosionsschutzverordnung – 11. ProdSV) in der Fassung der Bekanntmachung vom 12. Dezember 1996 (BGBl. I S. 1914), zuletzt geändert durch Artikel 21 des Gesetzes vom 8. November 2011 (BGBl. I S. 2178).
4. Verordnung über Sicherheit und Gesundheitsschutz bei der Bereitstellung von Arbeitsmitteln und deren Benutzung bei der Arbeit, über Sicherheit beim Betrieb überwachungsbedürftiger Anlagen und über die Organisation des betrieblichen Arbeitsschutzes, Betriebssicherheitsverordnung – BetrSichV (Operating Safety Regulation), als Artikel 1 der Verordnung vom 27.09.2002 zuletzt geändert durch Art. 5 des Gesetzes vom 08.11.2011 BGBl. I S. 2178.
5. Explosionsschutz-Regeln (EX-RL) Sammlung technischer Regeln für das Vermeiden der Gefahren durch explosionsfähige Atmosphäre mit Beispielsammlung zur Einteilung explosionsgefährdeter Bereiche in Zonen, BGR 104 Februar 2013.
6. EN 13 463–1:2009. (2009) *Non-Electrical Equipment for Potentially Explosive Atmospheres – Part 1: Basic Method and Requirements*.
7. Müller, R. (1977) Einfluss der Zündenergie auf die Zündgrenzen von Gas/Luft-Gemischen unter Variation von Druck und Temperatur. Dissertation an der Universität Erlangen-Nürnberg.
8. Pawel, D. and Brandes, E. (1998) *Abhängigkeit sicherheitstechnischer Kenngrößen vom Druck unterhalb des atmosphärischen Druckes*. Physikalisch-Technische Bundesanstalt (PTB), Braunschweig, September 1998.
9. EN 1839:2012. (2012) *Determination of Explosion Limits of Gases and Vapours*.
10. Nabert, K., Schön, G., and Redeker, T. (2004) *Sicherheitstechnische Kenngrößen brennbarer Gase und Dämpfe*, Band **I & II**, 3, erweiterte und vollständig überarbeitete Auflage, Deutscher Eichverlag.
11. EN ISO 1523:2002. (2002) *Determination of Flash Point – Closed Cup Equilibrium Method*.
12. EN ISO 2719:2002. (2002) *Determination of Flash Point – Pensky-Martens Closed Cup Method*.
13. Verordnung zum Schutz vor Gefahrstoffen (Gefahrstoffverordnung – GefStoffV) vom 26. November 2010 (BGBl. I S. 1643, 1644), zuletzt geändert durch Artikel 2 der Verordnung vom 15 Juli 2013 (BGBl. I S. 2514) am 1.12.2010 in Kraft getreten.
14. (1967) Council Directive 67/548/EEC on the approximate of laws, regulations and administrative provisions relating to the classification, packaging and labelling of hazardous substances of 27 June 1967.

Off. J., **196**, 1, last corrigendum by Regulation (EC) no. 1272/2008 (Official Journal no./s L 353/1), last adaptation by Directive 2000/2/EC of 15 January 2009 (Official Journal no./s L 11/6).
15. Regulation (EC) No 1272/2008 of the European Parliament and of the Council of 16 December 2008 on classification, labelling and packaging of substances and mixtures, amending and repealing Directives 67/548/EEC and 1999/45/EC, and amending Regulation (EC) No 1907/2006. *Off. J. Eur. Union* **L 353**, 51, ISSN: 1725–2555.
16. Steen, H. (2000) *Handbuch des Explosionsschutzes*, Wiley-VCH Verlag GmbH, Weinheim, New York, Chichester, Brisbane, Singapore, Toronto.
17. EN 14522:2005. (2005) *Determination of the Auto Ignition Temperature of Gases and Vapours*.
18. DIN 51794:2003. (2003) *Prüfung von Mineralölkohlenwasserstoffen – Bestimmung der Zündtemperatur (Testing of Mineral Oil Hydrocarbons – Determination of Ignition Temperature)*.
19. Gabel, D. (2000) Aufbau und Erprobung einer Messapparatur zur Bestimmung der Zündtemperatur in Abhängigkeit des Gemischdruckes in Anlehnung an DIN 51794 - Erste Messungen insbesondere von Stoffen mit niedriger Norm-Zündtemperatur. Diplomarbeit Technische Universität Bergakademie Freiberg, Fakultät: Maschinenbau, Verfahrens- und Energietechnik.
20. EN 60 079-20-1:2010. (2010) *Electrical Apparatus for Explosive Gas Atmospheres – Part 20: Data for Flammable Gases and Vapours, Relating to the Use of Electrical Apparatus*.
21. EN ISO 16852:2010. (2010) *Flame Arresters – Performance Requirements, Test Methods and Limits for Use*.
22. European Commission (2009) Council Directive 2009/104/EC of the European Parliament and the Council of 16 September 2009 on minimum safety and health requirements for the use of work equipment by workers at work (Second individual directive in the meaning of article 16, chapter 1 of Directive 89/391/EEC). *Off. J.*, **L260**, 5.
23. EN 1127–1:2011. (2011) *Explosive Atmospheres – Explosion Prevention and Protection – Part 1: Basic Concepts and Methodology*.
24. EN 13 463–8:2003. (2003) *Non-Electrical Equipment for Potentially Explosive Atmospheres – Part 8: Protection by Liquid Immersion "k"*.
25. EN 1012–2:1996+A1:2009. (2009) *Compressors and Vacuum Pumps – Safety Requirements – Part 2: Vacuum Pumps*.
26. European Commission (2006) Directive 2006/42/EC of the European Parliament and the Council of 17 May 2006 on machinery, and amending Directive 95/16/EC (revised version). *Off. J.*, **L157**, 24–86.
27. European Commission (2012) Directive 2004/108/EC of the European Parliament and the Council of 15 December 2004 on the approximation of the laws of the Member States relating to electromagnetic compatibility and for the annulment of the Directive 89/336/EEC. *Off. J. Eur. Union*, **C321**, 1.

Further Reading

Bartknecht, W. (1993) *Explosionsschutz- Grundlagen und Anwendungen*, Springer-Verlag, Berlin, Heidelberg, New York, London, Paris, Tokyo, Hong Kong, Barcelona, Budapest.

EN 13 463–2:2004. (2004) *Non-Electrical Equipment for Use in Potentially Explosive Atmospheres – Part 2: Protection by Flow Restricting Enclosure "fr"*.

EN 13 463–3:2005. (2005a) *Non-Electrical Equipment for Use in Potentially Explosive Atmospheres – Part 3: Protection by Flameproof Enclosure "d"*.

EN 13 463–6:2005. (2005b) *Non-Electrical Equipment for Use in Potentially Explosive Atmospheres – Part 6: Protection by Control of Ignition Source "b"*.

EN 60 079–0:2009. (2009) *Electrical Apparatus Intended for Use in Explosive Gas Atmospheres Part 0: General Requirements*.

EN 13 463–5:2011. (2011) *Non-Electrical Equipment Intended for Use in Potentially Explosive Atmospheres – Part 5: Protection by Constructional Safety "c"*.

7
Measurement Methods for Gross and Fine Vacuum

Werner Große Bley

Gases exert forces on the walls of a closed volume. These result from collisions of the moving molecules with the walls. Pressure is defined as the ratio of normal force F_N on a plane surface portion A. The resulting unit according to the SI system is:

$$1\ \text{Pa(Pascal)} = 1\ \text{N}/1\ \text{m}^2 (\text{Newton/square metre})$$

The sum of all molecular collision forces related to the unit area is called total pressure, the proportional pressure of a gas species is named 'partial pressure'. The sum of the partial pressures of all gas species present in the closed volume yields the total pressure (Dalton's law).

7.1
Pressure Units and Vacuum Ranges

According to the SI system the unit of pressure is 'Pascal'. Additionally the units 'bar' ($=10^5$ Pa) respectively 'mbar' ($=100$ Pa) are in use and accepted.

For a direct pressure measurement, the difference between the pressure to be measured and a known reference pressure is measured. Because of this measurement method engineers specify three different kinds of pressure (see DIN 1314):

- absolute pressure, measured against pressure 'Zero' of an empty space
- gauge pressure (positive or negative), measured against the local atmospheric pressure
- differential pressure, measured against a second pressure.

In vacuum technology normally the absolute pressure is stated. To generate 'Zero' pressure, a reference vacuum has to be generated with a pressure small compared to the pressure to be measured (there is no empty space in practice).

The old unit 'physical atmosphere' (atm) for absolute pressure was named after the scientist Torricelli and his evaluation of atmospheric pressure with a mercury vacuum gauge:

$$1\ \text{atm} = 760\ \text{Torr}(= 760\ \text{mm Hg}) = 101325\ \text{Pa}$$

Vacuum Technology in the Chemical Industry, First Edition. Edited by Wolfgang Jorisch.
© 2015 Wiley-VCH Verlag GmbH & Co. KGaA. Published 2015 by Wiley-VCH Verlag GmbH & Co. KGaA.

The units 'Torr' (1 mm Hg) and, derived from that, 'Micron' (1 µm Hg = 10^{-3} Torr) are still used in Anglo-Saxon countries. Here also the units 'psi' (also PSI or lb squin^{-1}, Pound-force per square inch, = 6894.8 Pa), 'psf' (=lb sqft pound-force^{-1} per square foot, = 47.88 Pa) and 'inch Hg' (=25.4 Torr) are in use. Another old unit is 1 dyn cm^{-2} = 0.1 Pa.

Besides the above-mentioned physical atmosphere which was only used for absolute pressure there was also the technical atmosphere 'at' used for absolute pressure (ata) or for gauge pressure (atü), overpressure referenced to atmospheric pressure. The unit 'atü' ('ü' for the German word 'Überdruck' = overpressure), as a result, had negative values in vacuum! The following relations hold between technical pressure units:

$$1 \text{ at} = 1 \text{ kp}/1 \text{ cm}^2 = 10 \text{ m WC (water column)} = 0.981 \text{ bar}$$
$$1 \text{ mm WC} = 9.81 \text{ Pa}; \ 1 \text{ p/mm}^2 = 1 \text{ m WS} = 98.1 \text{ mbar}$$

At last another unit no longer accepted should be mentioned, this is '% of vacuum'. In this case the local atmospheric pressure is set to 100%, too.

7.2
Directly and Indirectly Measuring Vacuum Gauges and Their Measurement Ranges

Pressure measurement in vacuum between 1000 and 10^{-12} mbar (15 decades!) requires different methods for the different pressure ranges. In general, one has to distinguish between vacuum gauges measuring directly and those measuring indirectly.

In directly measuring instruments, the pressure difference produces an elastic deformation or displacement of a separating member, for example a diaphragm. Indirectly measuring instruments measure a pressure proportional property (e.g. heat conductivity, ionized particle density or gas viscosity) and derive the pressure value from that.

In directly measuring vacuum gauges, the elastic restoring force of a spring element (diaphragm) or of gravitation (liquid column) is used for pressure measurement.

The scale of these instruments therefore is linear and the pressure measurement not gas-dependent. These instruments can only be used for the gross and fine vacuum range.

In indirectly measuring vacuum gauges (for lower pressures) the measured pressure proportional property is gas species dependent, too. That is why the displayed values are only correct for the gas species used when calibrating the instrument. Normally nitrogen (or air) is chosen for calibration and the resulting output values are called nitrogen equivalent pressure. For other gases, the output has to be corrected which is often neglected or forgotten in practice.

Because of the large measurement range of the indirectly measuring vacuum gauges (up to nine decades) the scale or electrical output of this type of instruments often has a logarithmic graduation, digital instruments show mantissa and exponent separately (e.g. 4.3×10^{-8} mbar). For further processing of measurement results, electrical instruments have analogue outputs (e.g. 0–10 V, 4–20 mA for recorder or controller units). Additionally, there are digital PC-interfaces or pressure-dependent trigger contacts, which are used for further evaluation or control.

In recent time so-called 'field-bus' interfaces (e.g. PROFIBUS, DeviceNet, LON) gain more importance. They may be used to integrate vacuum pressure transducers into complex process control systems without additional expense for cabling.

7.3
Hydrostatic Manometers

As the basic advantage of a hydrostatic manometer, the measurement result can be calculated from fundamental parameters making a calibration unnecessary. This has established their status of being the standards for calibration. Today these instruments are loosing their importance for every-day work since they are sometimes not very practical in use and if mercury is involved they pose an environmental hazard.

For a U tube manometer, which consists of a glass tube with constant diameter bent in the form of a 'U' filled with liquid, the pressure difference between the pressure to be measured and a reference pressure is determined using the following simple formula:

$$p = \rho \cdot g \cdot \Delta h$$

Here ρ is the density of the manometer liquid; g, the earth's gravitational acceleration and Δh, the difference of height of the two columns of liquid. If mercury at a temperature of 0 °C is used and Δh is measured in millimetre, pressure values can be stated in Torr (mm Hg). Inserting ρ, g and Δh in the SI-units kilogram, metre and second the result is given in the SI-pressure unit Pa.

In a U-tube manometer with one end closed, the saturated vapour pressure of the manometer liquid in the closed leg is used as the reference pressure. For mercury at room temperature, this pressure is approximately 2×10^{-3} mbar which is much smaller than the smallest pressure measurable (without additional effort) of approximately 1 mbar ($\Delta h = 0.75$ mm). Hence the measured pressure difference is taken directly as absolute pressure. The result is independent of gas species.

An extension of the measurement range of U-tube manometers was achieved by using the law of Boyle–Mariotte, which states a constant product $p \cdot V$ at constant temperature. In so-called compression vacuum gauges, therefore, the gas to be measured in the fine or high vacuum pressure range is compressed by a large factor (100 or more) so that again a visible deflection of the liquid column is generated. The compression factor results from the volume ratio of the measurement volumes. Well-known instruments are named after McLeod, Gaede and Kammerer. The latter is still in use in some places.

Besides the difficult handling another drawback of these instruments is condensation of vapours during measurement by the high compression factor. This leads to a lower pressure indication because not the total pressure but only the partial pressure of the non-condensated gases is measured

On the other hand just this effect is used to measure the so-called 'partial base pressure', of oil-sealed vacuum pumps. Here as a quality criterion the partial pressure of permanent gases (nitrogen, oxygen, etc.) is to be measured and the oil vapour pressure shall be neglected.

7.4
Mechanical and Electromechanical Vacuum Gauges

Similar to liquid manometers these measurement instruments use the force exerted on a separating element against a reference pressure. In mechanical gauges it is, however, a rigid diaphragm acting both for separation and as a spring with restoring force. Here also the gas or vapour species of the molecules have no influence since the deformation of the spring element is only proportional to the particle density and not their nature. The indication is effected by mechanical transformation of the deflection of the spring element to a scale. It is gas-independent.

The bourdon manometer (see also EN 837-1 [1]) consists of a tube (Cu/Be or stainless steel) bent in the form of a 'C' which is closed at one end and is rigidly fitted to a connection flange (Figure 7.1). Hence the pressure to be measured is inside while outside ambient air pressure acts as the reference. With overpressure the tube will straighten. If the pressure to be measured is lower than atmospheric pressure the flexion of the tube increases. The closed end of the bourdon tube is connected to a mechanics so that the deflection of the C-tube can be shown on a scale. One should be aware that the ambient air pressure used as a reference changes

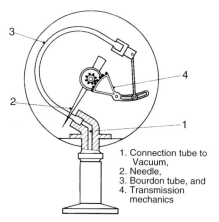

1. Connection tube to Vacuum,
2. Needle,
3. Bourdon tube, and
4. Transmission mechanics

Figure 7.1 Cut-away view of a Bourdon tube-vacuum gauge (measurement against ambient atmospheric pressure).

with a change of location (height) and weather (high/low pressure) and so can produce appreciable measurement errors for vacuum pressure measurements.

Diaphragm manometers (see also EN 837-3 [2]) use a planar or corrugated metal diaphragm, which show a pressure-dependent deflection used for pressure measurement. Diaphragm vacuum gauges can be made with a positive or negative gauge pressure scale. However, in contrast to bourdon vacuum gauges it is easily possible to produce absolute pressure instruments (with a reference vacuum) combining two convex diaphragms soldered under vacuum. These spring capsules are well-known from the normal barometer. To be used as a vacuum gauge they are installed in a vacuum-tight housing together with mechanics and a scale. When pressure is lowering in this housing the capsule will swell (its convexity will increase). The resulting translational movement is transferred by the mechanics and used as a measure of pressure change. A drawback is the immersion of the whole mechanics in the gas to be measured. Therefore, the use of these instruments is not recommendable with high amounts of dust or aggressive fluids present.

In a special diaphragm vacuum gauge (Figure 7.2), the mechanics is protected in the reference chamber which is separated by a stainless steel diaphragm. Additionally this instrument has a spreading of scale in the lower range resulting from a changing effective diaphragm area being very large at low pressure. At higher pressure the diaphragm is supported by a specific contour reducing its effective area. This design allows measurements in the whole pressure range of chemical engineering. Indications with nearly constant relative measurement uncertainty are possible in the range from 1 to 1000 mbar.

Normally the scale divisions of diaphragm vacuum gauges are spaced linearly and scales for positive and negative gauge pressure (planar spring) and absolute pressure (spring capsule) are possible. A statement of accuracy is made by a system of classes according to EN 837-3 [2] giving the amount of uncertainty as percentage of full-scale deflection.

Because of the spread of the scale the DIAVAC vacuum gauge in Figure 7.2 has a nearly constant measurement uncertainty of approximately 10% of measurement value in the whole range from 1 to 1000 mbar.

Sometimes diaphragm manometers for gauge pressure have an absolute pressure scale. Here one has to keep in mind that the measurement result is dependent on changes in atmospheric pressure. This effect can be partially compensated by rotating the scale or resetting the needle. For that purpose the instrument has to be evacuated down to a low pressure (lower than the lower limit of indication) and then to be adjusted to show zero.

7.4.1
Sensors with Strain Gauges

Diaphragm vacuum gauges where the deflection of the diaphragm is not detected mechanically but by strain gauges are called electromechanical transducers. In a strain gauge a thin element (wire or foil) is stressed by tension or pressure.

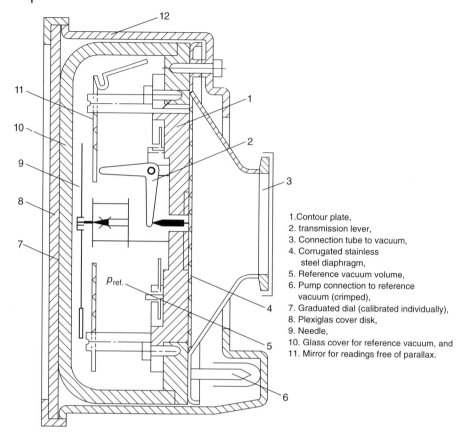

Figure 7.2 Cut-away view of a DIAVAC absolute pressure diaphragm gauge (manufacturer: Leybold).

This produces a change of electrical resistance within the element generated by the mechanical strain and not, as often assumed, by the change of length or cross section.

To achieve high sensitivity usually four strain gauges are combined to form a Wheatstone bridge with a large resistance change. The amplifier electronics is integrated within the transducer so that the signal is directly transformed into a standardized electrical signal (e.g. 4–20 mA). This unit is then called a pressure transducer or transmitter.

Piezo-resistive pressure sensors (Figure 7.3) make use of silicon membranes with integrated strain gauges that are generated in a semiconductor process. In this way low-cost sensors can be produced. These are not corrosion-resistant but they can be protected by a separating metal diaphragm.

Strain gauge or piezo-resistive pressure transducers can be bought as versions for absolute and gauge pressure. One disadvantage is the high temperature drift of the zero signal which constrains the measurement range to two pressure decades.

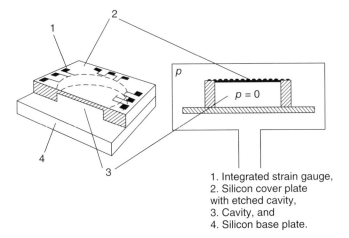

1. Integrated strain gauge,
2. Silicon cover plate with etched cavity,
3. Cavity, and
4. Silicon base plate.

Figure 7.3 Piezo-resistive absolute pressure transducer (principle).

In the capacitive diaphragm gauge (CDG, Figure 7.4) the diaphragm is part of a capacitor circuit. In CDGs a fixed capacitor electrode is installed opposite to the diaphragm and the diaphragm movement is captured by the change of capacitance during its pressure-dependent distortion.

Basically CDGs are classified according to the material of the measurement diaphragm. The standard configuration also used for high-end instruments contains a thin stainless (INCONEL) diaphragm or foil. Such extremely thin foils by nature have not a high resistivity against corrosion and also the overload tolerance of these sensitive sensors (down to less than 1 mbar full-scale deflection!) has to be observed.

For applications in corrosive fluids or at elevated temperature CDGs with ceramic diaphragm should preferably be used. Al_2O_3 is the best material for this application since it shows no creep after overloading and hence its zero stability is very good. Ceramic CDGs are nowadays available even for very low full-scale deflections below 1 mbar.

The capacitive measurement principle with its very high sensitivity allows a resolution of down to 10^{-5} of full scale. To minimize thermal errors, transducers with controlled temperature are available where the sensor cell is kept at a constant temperature (e.g. 45 °C) by electrical heating.

7.4.2
Thermal Conductivity Gauges

The heat conductivity of gases is independent of pressure as long as the distance of the two surfaces (or a filament to a cylindrical surface) with different temperature is much larger than the mean free path l of the gas molecules. Since l is inversely proportional to the pressure p the product $l \cdot p$ is constant. For air at 20 °C the

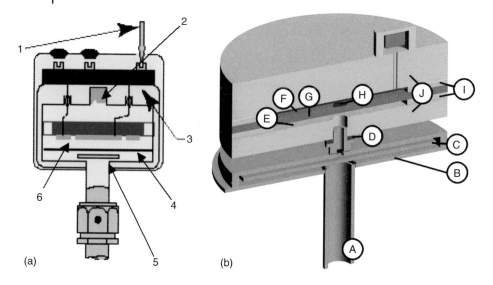

1. Zero adjustment screw,
2. Getter for reference vacuum,
3. Thermostat (optional),
4. Diaphragm (stainless steel),
5. Plasma shield (stainless steel),
6. Reference electrode (ring-shaped).

A. Vacuum connection (stainless steel),
B. Protective chamber (stainless steel),
C. Plasma shield (stainless steel),
D. Metal-ceramic connection (VACON),
E. Measurement chamber,
F. Reference chamber,
G. Diaphragm (Al_2O_3 >99.5%),
H. Electrode (gold),
I. Glass ceramic solder,
J. Sensor housing (Al_2O_3 >99.5%).

Figure 7.4 (a) Capacitive diaphragm gauge with metal diaphragm. (b) Capacitive diaphragm gauge with ceramic diaphragm.

product $l \cdot p = 6.5 \cdot 10^{-5}$ m·mbar. Hence at a pressure of 10 mbar the mean free path is approximately 0.65 mm.

In a thermal conductivity gauge (Figure 7.5) a heated filament about 7–30 µm in diameter is axially suspended in a measurement tube with an inner diameter of 10–20 mm. At 10 mbar with a mean free path l of 0.65 mm and a distance (from filament to wall) d of 5–10 mm the ratio of l to d is small enough so that the heat conductance by the gas from the heated filament to the wall (room temperature) is pressure dependent. This pressure-dependent heat conductance is the basis of the measurement principle.

Heat is not only conducted by the gas but also directly by the clamped ends of the filament and by radiation. These portions are not pressure dependent and lead to a residual indication, the so-called 'zero signal' or 'offset'. When the pressure has reached the respective offset signal, the heat conductance by the gas is small compared to the offset and it becomes difficult to be separated with sufficient accuracy. Therefore the pressure equivalent 'zero signal' determines the lower measurement limit of a heat conductivity gauge. Typically it is about 10^{-3} mbar.

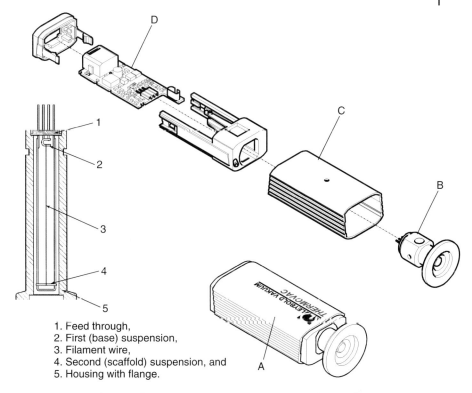

1. Feed through,
2. First (base) suspension,
3. Filament wire,
4. Second (scaffold) suspension, and
5. Housing with flange.

Figure 7.5 Pirani heat conductivity pressure gauge (cross sectional view of measurement cell). (A) Pirani transducer (sensor integrated with supply and interface electronics). (B) sensor housing with flange. (C) electronic housing. (D) PC-board (supply and interface).

7.4.3
Thermal Conductivity Gauges with Constant Filament Heating Power

In these vacuum gauges, the filament is heated with constant heating power resulting in a pressure-dependent specific temperature according to the dissipated heat at that pressure. The filament temperature can be measured by a thermoelectric element coupled to the filament. An easier way to determine the filament temperature is directly from its temperature-dependent change of electrical resistance (method of Pirani). The latter method is realized by making the filament one leg of a Wheatstone bridge that is adjusted to yield zero output voltage at lowest pressure (below the lower measurement limit).

With rising pressure the filament's temperature lowers and the resulting change in resistance detunes the bridge. This is shown by an indicating instrument that has an appropriate pressure scale.

The disadvantage of this very simple measurement device is the limited measurement range at high pressure (because of the low filament temperature) and the slow response to a sudden pressure rise (slow cool-down!).

7.4.4
Thermal Conductivity Gauges with Constant Filament Temperature

To eliminate the drawbacks of using different filament wires for different pressure ranges and to eliminate the slow response of the indication the Pirani principle was further developed. In an advanced gauge the filament is kept at a controlled constant temperature resulting in a constant resistance. This is achieved by adapting the bridge supply voltage to the heating power demand of the filament, keeping the bridge in a permanently balanced state.

The bridge voltage is then used to generate a pressure signal by means of an appropriate control amplifier circuit (a typical characteristic is shown in Figure 7.6). Modern units with microprocessor show the measurement value digitally with high resolution and generate an analogue output signal with precisely logarithmic pressure characteristic.

Thermal conductivity is not only dependent on pressure but also on gas species. Gases with small molar mass (Helium, Hydrogen) have a significantly higher heat conductivity than gases with higher mass (Argon, Krypton). Since the output of conductivity gauges is usually normalized to nitrogen or air a significantly higher indication for helium (at 15 mbar Helium pressure one gets a full-scale deflection of 1000 mbar!) and a significantly lower indication for Argon (e.g. 1000 mbar Argon yield only 40 mbar indication) has to be expected. The deviations are highest at pressures above 1 mbar, at pressures below 1 mbar the indicated and the actual pressure are proportional in first approximation. In this range a correction factor for different gases may be used (typical calibration characteristics are shown in Figure 7.7).

A better information is given by the correction tables issued by manufacturers, which take into account the geometry and the functional principle of their Pirani cells, too. For gas mixtures it is advisable to set up an individual correction table by comparative measurement against a gas-independent diaphragm gauge.

7.4.5
Environmental and Process Impacts on Thermal Conductivity Gauges

Small changes in the above-mentioned pressure independent offset can lead to appreciable measurement errors. Especially a good compensation of ambient temperature effects has to be provided to keep the specified measurement uncertainty in the operating temperature range. For this purpose a temperature-dependent resistor is often installed at the measurement tube.

An increase of heat radiation induced by blackening of the filament in the process environment can also lead to a considerable increase of zero signal. After a prolonged operation of a heat conductivity gauge it is not unusual that its offset rises by deposition of hydrocarbons on the filament resulting in a constant indication of 10^{-2} mbar with the real pressure being much lower.

In corrosive process fluids one has to use a corrosion resistive measurement cell. Such systems use nickel or platinum instead of tungsten as filament material,

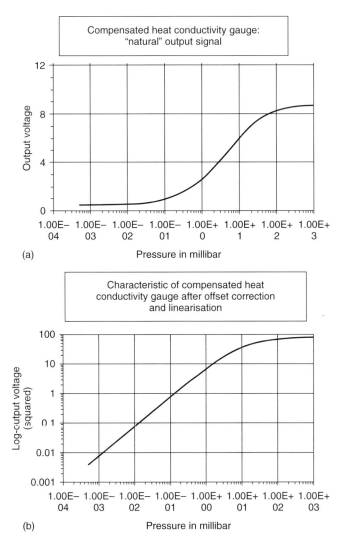

Figure 7.6 Output signal of a 'compensated' heat conductivity gauge (bridge voltage vs pressure) (a) pressure-dependent output voltage without correction and (b) linearization by logarithmic plot of squared output voltage (proportional to filament power) and subtraction of offset power (losses by radiation and heat conductance by suspension).

stainless steel instead of aluminium for the tube and alumina instead of glass as insulator material for the feed-through.

To protect the measurement cell against dust and oil mist from the process a filter made of sintered metal or steel wool is placed in front of the cell. It has to be exchanged regularly. The flow conductance must not be restricted to avoid measurement errors (obey manufacturer's instructions!).

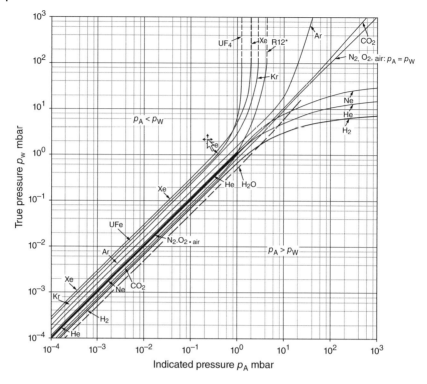

Figure 7.7 Calibration curves of heat conductivity gauge (p_A = indicated pressure and p_W = true pressure) for different gas species.

Zero signal drift may be compensated to a certain extent by trimming potentiometers installed either at the sensor head or the control unit. Potentiometers at the sensor head have the advantage that they can be preadjusted by the manufacturer so that each sensor immediately gives a correct indication when exchanged.

If the filament is heavily contaminated in chemical processes from the deposition of substances the cell may be simply cleaned by rinsing with solvents. Thus the distortion of the cells characteristic resulting from filament diameter change and/or discoloration can be corrected to a certain extent.

References

1. EN 837-1. (1996) *Pressure Gauges – Part 1: Bourdon Tube Pressure Gauges – Dimensions, Metrology, Requirements and Testing*, European Standards.
2. EN 837-3. (1996) *Pressure Gauges – Part 3: Diaphragm and Capsule Pressure Gauges – Dimensions, Metrology, Requirements and Testing*, European Standards.

Further Reading

EN 837-2. (1997) *Pressure Gauges – Part 2: Selection and Installation Recommendations for Pressure Gauges*, European Standards.

Lafferty, J.M. (1998) *Foundations of Vacuum Science and Technology*, John Wiley & Sons, Inc.

8
Leak Detection Methods

Werner Große Bley

8.1
Definition of Leakage Rates

It is reasonable to specify the tightness of systems and components by threshold values since absolute tightness cannot be achieved and is not necessary.

The leakage rate in chemical engineering is expressed as amount of substance lost in a given period of time with the unit kg h^{-1}.

However often the leakage rate is defined as volume throughput in a period of time with the unit mbar · l · s^{-1} (corresponds to cm^3 s^{-1} at 1 bar).

To convert the amount of substance into throughput rate the equation of ideal gases is used:

$$Q_{pV} = \frac{p \cdot V}{t} = \frac{m}{t} \cdot \frac{R \cdot T}{M} = Q_m \cdot \frac{R \cdot T}{M}$$

with

Q_{pV}	=	leakage rate as volume throughput rate
P	=	gas pressure (here air or helium)
V	=	transported gas volume
Q_m	=	leakage rate as mass flow
m	=	mass of the transported substance
M	=	molar mass of the specified gas
R	=	universal gas constant
T	=	absolute temperature (in K)
t	=	time.

Obviously the leakage rate Q_{pV} depends on the temperature and molar mass of the leaking gas. Therefore in a statement of a leakage rate in volume throughput units the gas species, for example air or nitrogen and a reference temperature, for example 23 °C must be stated additionally.

8.2
Acceptable Leakage Rate of Chemical Plants

A definition of the acceptable leakage rate of a chemical plant has to be guided by the degree of tightness that is required. Colloquial descriptions like 'liquid-tight' or 'gas-tight' have to be transformed into measurable values. As a guideline Table 8.1 can be used.

To define a tolerable singular leakage rate for a chemical process chamber the acceptable global leakage of a system has to be taken into account. It is difficult to state global leakage rates for air that can be used as a guideline for practical work since the requirements have to be defined according to the state-of-the-art and the process medium [1]. In any case it should be taken into account that the global leakage rates vs. system volume increases with an exponent of 2/3, since in

Table 8.1 Terms for threshold leakage rates.

Colloquial statement of tightness	Volume throughput leakage (mbar·l·s^{-1})	Mass flow leakage (kg h^{-1})	Explanations
Water tight	10^{-2}	10^{-5}	No water droplets dripping from leak (other liquids may still escape!)
Vapour-tight	10^{-3}	10^{-6}	No wetting of surrounding cold surfaces by escaping water vapour (still leaky for gases (!) but not enough vapour escaping to show visible condensation)
Bacteria tight	10^{-4}	10^{-7}	At this measurable leakage rate 50% of bacteria are retained (real bacteria tightness is only achievable with a completely dry leak channel)
Petrol/oil tight	10^{-5}	10^{-8}	In contrast to water, petrol and oil have much lower surface tension and/or viscosity and escape much easier through a small leak
Virus tight	10^{-6}	10^{-9}	At this measurable leakage rate 100% of bacteria and most viruses (in liquids) are retained
Gas-tight	$\ll 10^{-7}$	$\ll 10^{-10}$	There is no 'absolute' tightness, therefore a leakage rate has to be specified depending on application conditions
Detection limit for helium leak detector	10^{-10}	10^{-13}	Minimum detectable leakage rate under industrial conditions (under ideal conditions this limit may be lower by 1–2 orders of magnitude)

first approximation the number of leaks is proportional to the surface area of the system.

Small systems with a volume in the range of several cubic metres as used in pharmaceutical processes or vitamin production are usually tested down to a global leakage rate in the range of 10^{-4} mbar $\cdot$ l $\cdot$ s^{-1} (or ~10^{-3} g h^{-1} air). For tanks in petrochemical industry, leakages up to 10^{-1} mbar $\cdot$ l $\cdot$ s^{-1} (or ~1 g h^{-1} air) are acceptable. For a single leak, the leakage rate has to be smaller by 1 or 2 orders of magnitude, that is one will have to test for 10^{-6} mbar $\cdot$ l $\cdot$ s^{-1} for smaller systems with high quality processes but still for at least 10^{-3} mbar $\cdot$ l $\cdot$ s^{-1} ('water vapour-tight') for large tanks. As one can see from Table 8.1 this means that 'real' tightness for petrol or oil is not yet achieved for a single leak ($<10^{-5}$ mbar $\cdot$ l $\cdot$ s^{-1} necessary!).

8.3
Methods of Leak Detection

A review of the common leak detection methods in chemical engineering is given in Table 8.2.

Table 8.2 Leak detection methods and their characteristics.

Method	Test fluid	Threshold leakage rate (mbar $\cdot$ l $\cdot$ s^{-1})	System pressure	Leak localisation	Quantitative leakage rate statement
Pressure rise	Air or process gas	10^{-4}	Vacuum	No	With large measurement uncertainty
Pressure drop	Air or process gas	10^{-4}	Overpressure	No	With large measurement uncertainty
Bubble test with foam	Air or process gas	10^{-5}	Overpressure	Yes	No
Bubble test in water bath	Air or process gas	10^{-4}	Overpressure	Yes	With large measurement uncertainty
Ultrasonic detection	Air or process gas	10^{-2}	Overpressure	Uncertain	No
Heat conductivity detection	Air or process gas	10^{-4}	Overpressure or vacuum	Yes	No
Mass spectrometer helium leak detector	Helium	10^{-7}	Overpressure (sniffing)	Yes	With traceable calibration according to requirements in ISO 9000
		10^{-11}	Vacuum		

The presented methods can be divided basically into two categories:

1) pressure change methods (without specific tracer gas)
2) tracer gas methods.

Additionally one has to distinguish by direction of gas flow:

a. gas flow out of system (typically overpressure in the system)
b. gas flow into system (typically vacuum in the system).

A guide to select appropriate test methods can be found in the European standard EN1779 (a description of pressure change methods and tracer gas methods can be found in the standards EN 13184 and EN 13185, respectively).

From the table one can see that only helium leak detection with a mass spectrometer leak detector allows a real quantitative statement of leakage rates compliant with the requirements of the ISO 9000 series of standards for quality management. Therefore this method and its variants will be described in more detail in the following.

8.4
Helium as a Tracer Gas

Helium which is mostly used as a tracer gas in leak detection has excellent properties for tightness tests of chemical systems.

- It is non-poisonous.
- It is non-inflammable.
- It behaves inert, does not react with other substances.
- It is stable at higher temperature, can be used in hot processes.
- It is non-condensable in the whole technical application range.
- Only 5 ppm are found in natural air, hence detection of smallest leaks possible.
- It shows no interference when detected with a mass spectrometer.

8.5
Leak Detection with Helium Leak Detector

No other known leak detection method achieves the detection capability and certainty as the helium leak detector. As operational conditions in chemical engineering can vary appreciably the job of testing systems or finding leaks cannot be done using a single approach. Normally helium leak detectors are used only to test vacuum systems as the cost of the tracer gas for leak checking atmospheric or overpressure systems would be too high. The best test conditions for vacuum systems are to be found during production, that is during the time when the process supports the testing, for example by the existing vacuum pumps. Leak detectors cannot be directly connected to systems in areas with explosion hazards as they are not produced in explosion-proof versions. A work-around will be described later.

Before connecting a helium leak detector to a system and choosing accessories the following questions should be answered.

- How low is the operating pressure of the vacuum system?
- Are there solid or liquid, aggressive or toxic substances which can reach the leak detector?

The working method is strongly dependent on the leak detector connection: in fine vacuum the detector can be connected directly whereas in gross vacuum throttles, leak valves, traps, filters or cold-traps have to be inserted.

8.6
Leak Detection of Systems in the Medium-Vacuum Range

8.6.1
Connection of Leak Detector to the Vacuum System of a Plant

If the system pressure is below the maximum tolerable inlet pressure of the leak detector (e.g. 0.2 mbar, with newer instruments even up to 5 or 10 mbar) the helium leak detector can directly be connected to the system.

In all other systems with operating pressure too high for the tolerable pressure of direct leak detector connection the helium leak detector can be connected via an inlet pressure adaptation with a throttling valve. The throttling valve has to be protected by a small condensation trap since liquid in the pump line has to be expected. Connection to systems with aggressive or dangerous products is more expensive. Here ingress of product has to be avoided by any means. This is accomplished by small cold-traps filled with liquid nitrogen. At high product flow and long exposure it may be necessary to use two alternating traps.

The real detection sensitivity and the transfer time of the gas from the leak to the helium detector are decisive for the performance of leak detection. The effective sensitivity is mainly given by the pumping speed of the system vacuum pump in relation to the pumping speed of the helium leak detector. Since partial flow factors between 10^3 and 10^4 are not untypical in practice, helium leak detectors with a typical detection limit of better than 10^{-10} mbar $\cdot$ l $\cdot$ s^{-1} can take advantage of their excess sensitivity. In principle, even with only 1000 of the leakage rate entering the leak detector (partial flow factor 1000) air leakages down to 10^{-9} kg h^{-1} can be detected. This makes sense for chemical production since normally single leaks have to be found.

In Figure 8.1, the schematics of a distillation system with high, medium and rough vacuum pumps is shown. If leak detection has to be performed on such a system the question arises at which position best to connect the leak detector. The possible connection points are designated with 1–4. Criteria for the optimum connection point are the minimum leakage rate to be detected and the prevailing pressure. For multistage pumping systems there is a general rule: the closer the

8 Leak Detection Methods

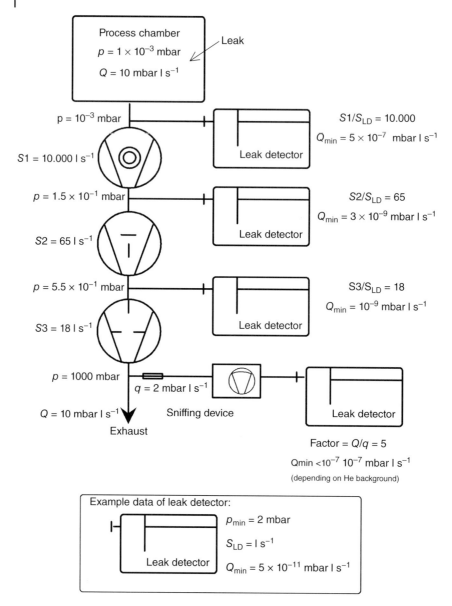

Figure 8.1 Vacuum system with high, medium and rough vacuum pumps: different connection positions for the helium leak detector.

connection point is to the process vacuum the lower the pressure (so no throttle is necessary), but on the other hand the higher is the pumping speed of the installed system vacuum pump. This results in a higher (worse) lowest detectable leakage rate.

8.6.2
Detection Limit for Leakage Rates at Different Connection Positions of a Multistage Pumping System

The detectable threshold leakage rate Q_{thresh} results from the ratio of pumping speeds of the respective system vacuum pump S_{Pump} and the helium leak detector pump S_{LD} (partial flow factor) multiplied by the minimum detectable leakage rate $Q_{LD,min}$ of the helium leak detector:

$$Q_{thresh} = \frac{S_{Pump}}{S_{LD}} \cdot Q_{LD,min} = \gamma \cdot Q_{LD,min}$$

where γ is the partial flow factor.

This dependency is shown for a leak detector connection point above S1:

$$Q_{thresh} = \frac{10.000 \ l \times s^{-1}}{0.2 \ l \times s^{-1}} \times 5 \cdot 10^{-11} \ \text{mbar} \times l \times s^{-1} = 5 \cdot 10^{-7} \ \text{mbar} \times l \times s^{-1}$$

Hence if a threshold leakage rate for a single leak of maximum $1 \cdot 10^{-7}$ mbar $\cdot$ l $\cdot$ s^{-1} is to be tested this connection point cannot be chosen. Looking at Figure 8.2, one can see that a connection point above S2 is much better since threshold leakage rates for a single leak down to $3 \cdot 10^{-9}$ mbar $\cdot$ l $\cdot$ s^{-1} can be detected. At the connection point above S3 even a leakage rate of $1 \cdot 10^{-9}$ mbar $\cdot$ l $\cdot$ s^{-1} can be detected and

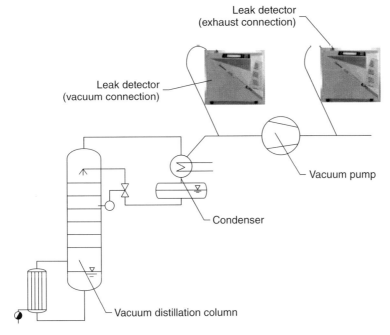

Figure 8.2 Vacuum distillation system with a helium counter-flow leak detector connected.

even at the higher pressure of $6 \cdot 10^{-1}$ mbar the helium leak detector can be connected without a throttling valve for inlet pressure reduction (instruments with lower tolerable inlet pressure as assumed here will however need such a throttle!).

8.7
Leak Detection on Systems in the Rough Vacuum Range

For the connection of leak detectors to systems operating in the rough vacuum range there are several possibilities with specific advantages and disadvantages.

8.7.1
Connection of Leak Detector Directly to the Process Vacuum

Vacuum systems in the rough vacuum range can also be tested successfully with helium leak detectors in all ranges of rough vacuum pressures. A large reduction of sensitivity because of the unfavourable partial flow factor is no problem since in those systems fine leaks normally are not relevant. To connect the helium leak detector (optimum working pressure <2 mbar) to the system (operating pressure typically 10–100 mbar) a throttling technique is used. The helium leak detector is connected to the system as shown in Figure 8.2. The throttling means is inserted at the connection point of the system. It can be a leak valve (needle valve), a ceramic or glass frit or a selectively permeable plastic diaphragm. The leak valve is recommendable in the case of varying system pressure whereas the other throttles are a convenient and safe solution when pressure changes are small.

8.7.2
Connection of Leak Detector at the Exhaust of the Vacuum System

A specific option of leak testing vacuum systems is the use of a helium sniffer in the exhaust line of the process vacuum pump. The helium sniffer is an option to the helium leak detector and allows admission of small amounts of gas from atmospheric pressure into the operating vacuum of the leak detector via appropriate pressure reducing throttle elements. During exhaust leak detection, the sniffer tip is inserted into the exhaust line via a small bore hole (see Figures 8.1 and 8.2). In this way the helium concentration within the exhaust line is measured and monitored continuously. Helium ingressing through a leak will be pumped by the vacuum pump and is transferred into the exhaust line together will all other gases. Concentration changes >5 ppm can be detected reliably by this arrangement. However one will have to care that no air from ambient is sucked in (e.g. when the sniffer tip is close to the exhaust opening).

With this arrangement a leak test with helium can be performed even on systems in areas with explosion hazards since sniffing lines with a length of 50–60 m allow a set-up of the leak detector outside the hazardous area. However in using such long tubes a dead time of several seconds has to be expected until a signal can be observed.

8.8
Leak Detection and Signal Response Time

The signal response time is important for the safe localization of leaks. It states the time period after spraying helium on the leak until the leak detector indication reaches a defined level (63% of final equilibrium value). Figure 8.3 shows the typical exponential trace of the signal rise. The response time can be calculated from the parameters of the system under test. The critical factors are the volume of the system and the total pumping speed of the installed vacuum pumps. Typical response times of chemical plant systems are in the range of several seconds up to a few minutes.

As the rising signal follows an exponential function and the 100% signal is only reached after indefinitely long time it is reasonable to use the time periods for a 63 and 95% signal as a characteristic. The 63% time is given by

$$\tau = \frac{V_{system}}{S_{system}}$$

with

V_{system}	–	volume of system
S_{system}	=	total pumping speed of vacuum pumps.

To achieve 95% of the final signal value one will have to wait for the time $t_{95} = 3 \times \tau$.

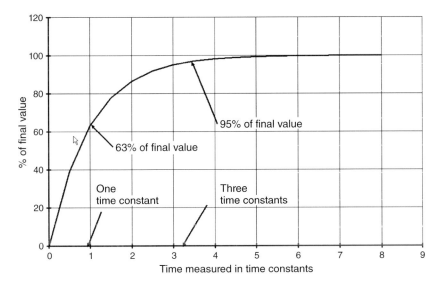

Figure 8.3 Exponential trace of a leak rate indication after spraying helium on a leak position.

A tolerable signal response time can only be achieved with a large pumping speed. With pumping speed being too small only large leakages can be localised since only the rather small signal portion in the steep rising part of the response curve is observed and the final value is far from being reached. By connecting a test leak to the system under test one can obtain reliable information about the reduced sensitivity and the signal response time. Preferably the connection point should be far away from the system pump to see realistic response times. Helium test leaks with a shut valve and with minimum dead volume are most suitable. To avoid any false reading of response time it is a proven procedure to leave the valve of the test leak open until an equilibrium value is reached, then close the leak as quickly as possible and measure the time until a decrease by 63% is reached.

8.9
Properties and Specifications of Helium Leak Detectors

In principle, a helium leak detector detects the partial pressure of a tracer gas in vacuum with a suitable sensor. Small mass spectrometers have become common as a sensor, especially small magnetic sector-field mass spectrometers with permanent magnets. Also the quadrupole mass spectrometer well known from residual gas analysis can be found in some leak detectors.

A mass spectrometer ionises all gas particles by means of electrons emitted from an incandescent filament. In a next step helium ions are filtered by a (magnetic or electric) separation system and are fed to an ion collector, which is connected to a sensitive current amplifier that can measure and indicate the extremely tiny electrical current.

A mass spectrometer is a partial pressure sensor that cannot directly output the leakage rate of a system. Although the partial pressure is proportional to the leakage rate it is also dependent on the pumping speed of the leak detector and the pumps installed on the system. To generate a leak detector from a mass spectrometer a high vacuum system, some valves and a control module have to be added. The resulting system is then capable of measuring leakage rates, that is gas throughput. The vacuum schematics of a typical helium leak detector is shown in Figure 8.4.

The main specifications of a modern general-purpose leak detector are the following (the given data are typical for a good leak detector):

- *Detection limit:* $\leq 5 \times 10^{-11}$ mbar $\cdot$ l $\cdot$ s^{-1} at ≈ 1 s time constant
- *Helium pumping speed:* ≥ 1 l s^{-1}
- *Maximum inlet pressure:* ≥ 5 mbar.

Definitions of these specifications of helium leak detectors are given in Table 8.3.

In recent years so-called 'dry' helium leak detectors have been developed. They do no longer contain any oil-sealed vacuum pumps but instead use diaphragm or scroll vacuum pumps sealed by elastomers in a 'dry' manner. Although these

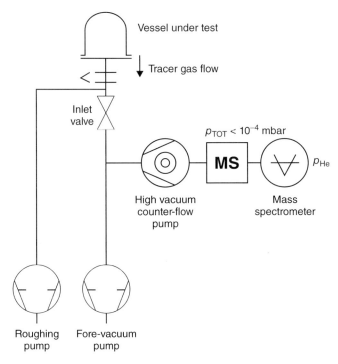

Figure 8.4 Vacuum schematics of a typical mass spectrometer helium leak detector (counter-flow principle).

leak detectors were developed for semiconductor processes they show a number of advantages for chemical engineering as follows:

- No adverse effects on processes when leak detector is faulty operated.
- No chemical reaction of process media with pump oil of leak detector.
- No frequent change of contaminated oil resulting in problems of disposal.

However, one has to consider the limited lifetime and chemical compatibility of gaskets and diaphragms of the dry pumps before they are used in specific processes.

8.10 Helium Leak Detection in Industrial Rough Vacuum Applications without Need of a Mass Spectrometer

The detection opportunities for leaks using a leak detector, which works with a mass spectrometer for the helium detection, are unmatched. However, there are also known disadvantages of this method.

In particular, when the working pressure in the vacuum system to be tested exceeds the acceptable total pressure at the leak detector inlet flange, difficulties

Table 8.3 Specifications of helium leak detectors.

Specification	Unit	Description
Detection limit (minimum detectable leakage rate). Makes only sense together with a statement of time constant!	*millibar × litre/second* $(mbar \cdot l \cdot s^{-1})$ or *pascal × cubicmetre/second* $(Pa \cdot m^3 \, s^{-1})$	This value states the smallest helium gas flow that can be distinguished safely from background noise (see also standards EN 1518 and ISO 3530)
		The longer the output signal is averaged the more favourable this value appears but the slower the response of the leak detector
Time constant of output signal	*second* (s)	This value states the typical rise time of the output signal (more precisely: the time during which the output signal rises to 63% of its final value after a step change of the leakage rate)
		This values gives a measure of the response time of the instrument
Helium pumping speed at test port	*litre/second* $(l\,s^{-1})$	This pumping speed determines the detection limit together with the system pumps (see 'partial flow factor' in the chapter about leak detection methods)
Maximum tolerable inlet pressure	*millibar* or *pascal* (*mbar*) or (*Pa*)	This pressure determines the type of system to which the leak detector can be connected without throttling
		For safe operation the maximum tolerable inlet pressure must be higher than the actual system pressure

may occur. In these cases, the total pressure must be reduced at the leak detector inlet with a throttle (e.g. aperture or glass frit). The throttle at the inlet flange of the leak detector reduces the gas conductance to the leak detector and therefore the working pressure at the leak detection inlet to an acceptable low pressure. Thus, the partial flow ratio (ratio of the pumping speed of the vacuum pump system and conductance to the leak detector) is increased, so the detection sensitivity of the helium leak detector is reduced proportionally.

Furthermore, there is a risk in operating a leak detector on the basis of a mass spectrometer at systems with condensable product. The conductance of the throttle could be changed or the inlet point clogged even completely due to condensed product. This means that the helium leak detector is getting blind to the test gas.

As already discussed above, an ion-separating system (ion spectrometer) is needed for the mass spectrometric detection of helium, which requires high vacuum condition. These systems operate at a total pressure of less than 10^{-4} mbar. For this purpose, a turbo molecular pump is applied in the leak detector. So-called hot cathode filaments are used in the separation system for ion generation. These filaments wear, depending on use conditions more or less quickly, and must be replaced after a possible unpredictable burnout.

The helium detection without using expensive high vacuum and without critical components is desirable for leak testing of industrial plants. In recent years a new type of helium sensor was therefore developed, which can detect helium selective and sensitive, without requiring high vacuum conditions and without the need of hot cathode filaments. It is realised a simple and reliable helium detector with this so-called Wise Technology® sensor, which is particularly suitable for industrial plants.

8.10.1
Principle of the Wise Technology® Sensor

Details to the basic principle of the Wise Technology® sensor can be found in the literature [2].

In the classical mass-spectroscopic detection, the entirety of the gas molecules and atoms of to be analysed atmosphere is first ionised and accelerated. It is done under high vacuum conditions. The ions are separated according to their mass/charge ratio in the magnetic sector field and subsequently detected.

The idea of the novel helium detection is a helium partial pressure sensor based on the known method of separating gases by a semipermeable membrane with gas-specific permeation. The component of the gas which is to be detected, in this case helium, is separated in the non-ionised state using the membrane. The amount of the separated helium is measured behind the membrane in a hermetically sealed volume, using a total pressure measurement in the evacuated volume. The change of total pressure in the enclosed volume corresponds to the helium partial pressure in front of the sensor membrane with a high selectivity for helium.

Quartz is well permeable to helium. The selectivity of a pinhole-free, amorphous quartz membrane is very high for helium. Only hydrogen and neon show a significant permeation rate, but for this application this is negligible. In the enclosed volume behind the membrane the helium partial pressure rises to a pressure proportional to the partial pressure of the process side environment of the sensor. The total pressure of the external environment of the sensor system does not show any influence on the pressure inside the sensor volume.

The permeability of quartz for helium, and therefore the sensitivity of the sensor depends on the temperature. For increasing the sensitivity and response time of the sensor, the quartz membrane is heated to about 350 °C during operation of the sensor.

8.10.2
Application

In an industrial plant with a multistage vacuum pump system (see the comments in the article above), the helium detector is connected in the same position as a mass spectrometer leak detector. The position with the lowest pumping speed in the vacuum pumping system is the most suitable location to connect the detector. It is the position with the highest total pressure and hence also with the highest partial pressure of the test gas. This location is usually the inlet flange of the fore-vacuum pump of the multistage pumping system. A device used for mobile helium leak detection with the novel Wise Technology® sensor, which works without high vacuum requirements, is described in [3]. This unit has been on the market since 2012. The leak tester is connected directly to the system to be tested for leaks at a total pressure of less than 200 mbar. It operates without a vacuum pump of its own, so the working pressure in the sensor is identical to the total pressure at the inlet flange.

This means the same test gas pressure in the sensor as in the pipes of the system to be tested; therefore no partial flow factor has to be considered. At high pressure a slight delay of the signal rise is observed, which is caused by the necessary diffusion of the test gas from the area of the connecting flange towards the sensor. However, when testing large systems the total reaction time is given by the time which the test gas needs to flow along the tubes from the leak position to the sensor location. This time delay, as well as the detection sensitivity of the system, is determined during the application by a specific calibration leak. A particular advantage of this solution is that the sensor information is transmitted directly to a wireless remote control. Thus, a technician can check components of large systems that are located far away from the sensor position. He sprays helium at the components and receives the signal via the remote control, without visual contact to the sensor.

A typical example of the application of the Wise Technology® sensor is the leak test of heat exchangers in power plants. On the low pressure side the steam turbine has vacuum conditions below $\ll 100$ mbar to increase the condensation of steam and thus optimise the efficiency. Therefore, the efficiency of the power plant is the better the lower the working pressure in the steam turbine condenser. The achievable pressure depends on the cooling water temperature and the overall tightness of the system. The integral tightness of the system and the total leakage rate in the condenser circuit can be determined by measurement of various parameters: the total pressure in the system or the mass flow at the outlet of the fore-vacuum pump or the gas concentration of components of air (e.g. oxygen) in the exhaust stream. In order to localise a leakage rate normally the classical helium leak detection method is used (helium leak detection with mass spectrometer and high vacuum system). As a new and very elegant method Wise Technology® comes on the scene. The sensor can be mounted directly at the system; both high working pressure and high humidity in the system are compatible with the requirements of this leak detection system. Other applications can be found in plants of the chemical industry with process steps under vacuum conditions.

References

1. Kistenbrügger, L. (1987) Emissionen aus Vakuumanlagen in der chemischen Technik. *Swiss Chem.*, **9** (3), 663.
2. Wetzig, D. (2009) Heliumpartialdrucksensor statt Massenspektrometer im Helium-Leckdetektor. *VIP Vakuum in Forschung und Praxis*, **21** (2), 8–13.
3. Pfeiffer, V. (2012) *MiniTest 300, Portable Leak Detector*, Pfeiffer Vacuum GmbH.

Further Reading

European Standards

Jousten, K. (ed) (2013) *Wutz Handbuch der Vakuumtechnik*, 11. Auflage, Vieweg

Patrick, O.M. (ed) (1998) *Nondestructive Testing Handbook: Leak Testing*, 3rd edn, vol. 1, American Society for Nondestructive Testing.

Dushman, S. and Lafferty, J.M. (eds) (1998) *Foundations of Vacuum Science and Technology*, 3rd edn, John Wiley & Sons, Inc.

EN 1330-8. (1998) *Non-Destructive Testing – Terminology-Part 8 Terms Used in Leak Tightness Testing*, European Committee for Standardization, Brussels.

EN 1779. (1999) *Non-Destructive Testing – Leak Test: Criteria for Method and Technique Selection*, European Standards.

EN 13184. (2001) *Non-Destructive Testing – Leak Test: Pressure Change Method*, European Standards.

EN 13185. (2001) *Non-Destructive Testing – Leak Test: Tracer Gas Method*, European Standards.

EN 13625. (2001) *Non-Destructive Testing – Leak Test: Guide to the Selection of Instrumentation for the Measurement of Gas Leakage*, European Standards.

9
Vacuum Crystallisation
Guenter Hofmann

9.1
Introduction

Crystallisation is one of the oldest unit operations known to mankind. Even today solar evaporation is still customary in regions with plentiful sunshine. Advantages are the very simple technology and a cost-free supply of energy. On the other hand, a very large surface area is needed. The rates of production per unit surface area fall in the range of $3-10\,\mathrm{g\,m^{-2}h^{-1}}$. The purity of the product which can be attained in the production of salt in solar ponds is also limited. The product always consists of aggregates of single crystals which have grown together, this leading to inclusions of mother liquor and higher residual moisture. Untreated salt from solar ponds can be obtained with purities up to about 98% (Figure 9.1).

In modern equipment for crystallisation, on the other hand, any crystalline product can be produced with purities of about 99.9% and with almost any desired crystal size distribution (CSD). The specific rate of production in industrial crystallisers is several thousand times larger than in solar ponds [1–4].

9.2
Crystallisation Theory for Practice

The result of any crystallisation process is a crystalline matter with a certain CSD, certain crystal habit and purity. These properties very often are quality requirements and well defined by market demands. Certainly, also the crystallisation process itself requires a minimum CSD, as the resulting suspension still has to be separated. This separation can be effected the more perfectly (purity) as more compact (crystal habit) and coarser (CSD) the crystals are. Quality and economy of the separation process crystallisation are therefore strongly dependent from crystal size, CSD and crystal habit. These properties also take influence to the storage ability, the dust-freeness and also the bulk density of the product.

Vacuum Technology in the Chemical Industry, First Edition. Edited by Wolfgang Jorisch.
© 2015 Wiley-VCH Verlag GmbH & Co. KGaA. Published 2015 by Wiley-VCH Verlag GmbH & Co. KGaA.

Figure 9.1 Solar pond salt (GEA Messo).

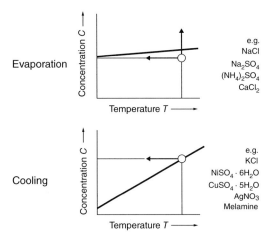

Figure 9.2 Crystallisation processes, supersaturation via cooling or evaporation (GEA Messo).

The key for complying all these demands is given already by simple theoretical considerations [5]. The most important factor is appropriate handling of supersaturation, driving force of crystallisation. Supersaturations can be produced by evaporation or cooling (Figure 9.2). In case of solubilities with low dependencies on temperature the evaporation is the normal choice, whereas with strong temperature dependencies the cooling method is more attractive.

As higher the supersaturations are, as faster the crystal growth is and as more effective the crystalliser. Certainly, not any supersaturation can be chosen, because also the nucleation processes are supersaturation dependent. There is the spontaneous or primary nucleation which is caused by a critical height of supersaturation

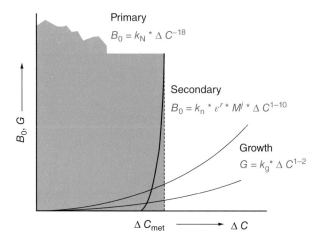

Figure 9.3 Kinetics (GEA Messo).

In the figure:
- Primary: $B_0 = k_N * \Delta C^{\sim 18}$
- Secondary: $B_0 = k_n * \varepsilon^r * M^j * \Delta C^{1-10}$
- Growth: $G = k_g * \Delta C^{1-2}$

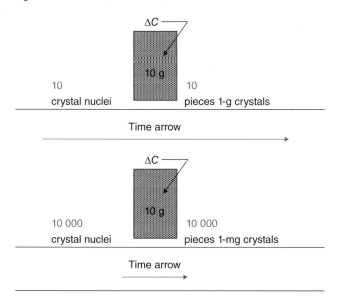

Figure 9.4 Influence of nucleation on crystal size (GEA Messo).

and the secondary nucleation which is depending on the presence of crystals and appears already lower supersaturations within the metastable field. This nucleation mechanism is fed by crystal/crystal impacts and by impacts with parts of the crystalliser and therefore can be controlled by the energy input (Figure 9.3).

As smaller the nucleation rate is, as coarser the single crystal will be. Figure 9.4 shows a graphical explanation. Following a production capacity of 10 g, 10 pieces of 1 g crystals are generated if there are 10 nuclei only. In comparison there will result 1 mg crystals, only, if the number of crystals is brought to 10 000 instead of

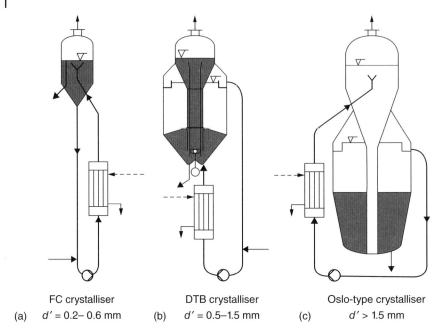

	FC crystalliser		DTB crystalliser		Oslo-type crystalliser
(a)	$d' = 0.2–0.6$ mm	(b)	$d' = 0.5–1.5$ mm	(c)	$d' > 1.5$ mm

Figure 9.5 (a–c) Basic types of crystallisers (GEA Messo).

10. Because supersaturation cannot be set to any height, significantly more time is necessary to crystallise 10 g crystals than 1 mg crystals. This relationship is indicated by the time arrows in Figure 9.3 and makes clear the strong effort necessary when coarse instead of fine crystals shall be produced. For the design of crystallisers that leads to the following most important design rules:

- Any supersaturation in crystallisers must be smaller than the metastable range.
- Supersaturations have to be chosen high for efficient crystal growth rates.
- Secondary nucleation has to be controlled by the input of mechanical energy.

Crystallisers designed according to these rules will certainly come to function. Of course, these principles can be found in the well-known basic types of crystallisers (Figure 9.5a–c).

A typical design feature is the way supersaturation is controlled and kept within the metastable range. The method is independently from the crystallisation process. Figure 9.6 explains this method based on the vacuum cooling crystallisation principle. This principle is also valid for the vacuum evaporation, surface cooling and reaction crystallisation.

On the left hand side, it is shown a sketch of a (forced circulated) FC-type crystalliser, a crystalliser with external circulation loop. On the right hand side, one can find a simplified solubility system in which the metastable limit is marked with a dashed line. What happens within one circulation loop is indicated by numbers in the solubility diagram as well as in the FC-crystalliser sketch.

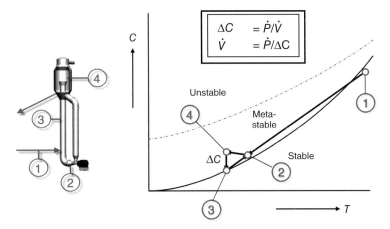

Figure 9.6 Control of supersaturation (GEA Messo).

The suspension (3) in the crystalliser is mixed with feed liquor (1). In vacuum-cooling crystallisation this feed is hotter and more concentrated than the circulated suspension. As consequence, the mixture (2) is more concentrated and hotter than the original suspension. The mixture is pumped to the evaporator vessel at the top of the crystalliser (4). There, a constant solvent partial pressure is kept constant by controlled condensation of the evaporated solvent. The solution starts boiling and gets re-cooled adiabatically thus supersaturating the solution, supersaturation being represented by the line 3–4. The generated ΔC is consumed by crystal growth.

With entering point 3 again the circulation loop gets closed. The height of the generated supersaturation (4) can be adjusted by varying the circulation flux through the crystalliser. With more solution circulated the points 2 and 4 come nearer to point 3. The definition figures of point 1, that is the feed mass flow and the crystalliser cooling (operating) temperature define the production capacity of the system. The circulation flow in the crystalliser is chosen adapted to this production capacity and by this way the tip supersaturation is kept within the metastable range. The circulation flow, therefore, is a major design criterion and for a certain product and a certain production capacity a pre-condition for all types of crystallisers. The circulation flow can be calculated by Eq. (9.1).

$$\Delta C_{max} = 0.5\,\Delta C_{met}; dV/dt = \frac{dP/dt}{(0.5\,\Delta C_{met})} \quad (9.1)$$

This relation is limiting the production capacity of any crystalliser. Essential increases for existing crystallisers are not permitted, otherwise spontaneous nucleation can occur. Comparably high circulation flows are necessary already for minor production capacities, because the metastable fields usually are in the range of some grams per litre only. For example, for a production of $1\,t\,h^{-1}$ and a ΔC of $1\,g\,l^{-1}$ the circulation flow already comes up to $1000\,m^3\,h^{-1}$.

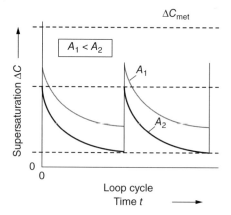

Figure 9.7 Supersaturation cycles (GEA Messo).

In Figure 9.7, this is repeated for one further design criterion which was neglected above, the dependency of desupersaturation on the suspended crystal surface area. After the production of supersaturation there will be crystal growth on the crystal surface area (A_1) while the suspension is circulated through the crystalliser. By the sequences of circulation loops always a typical saw tooth function will be formed. Because the remaining supersaturation gets added to the supersaturation generated in the next loop, the supersaturation has to be consumed preferably within a loop. With more crystal surface area offered (A_2), the desupersaturation rate becomes faster, resulting in lower supersaturations remaining after the loop. Suspension densities (mass of crystals/mass of suspension) between 15 and 25 wt%, therefore, are in practical use.

The mass deposition rate dm/dt (Eq. (9.2)) can be described as exponential function with supersaturation ΔC as exponent basis and the crystal surface A as linear factor. The proportionality factor k_g considers the influence of the temperature.

$$\frac{dm}{dt} = k_g \cdot A \cdot \Delta C^m = -\frac{d(\Delta C)}{dt} \tag{9.2}$$

Also the nucleation rate B_0 (Eq. (9.3)) can be described as exponential function with supersaturation as exponent basis and the specific energy input ϵ and the crystal mass m_T as additional exponential factors.

$$B_0 = k_N \cdot \epsilon^r \cdot m_T^i \cdot \Delta C^n \tag{9.3}$$

As explained above two of the main influence factors are fixed for ensuring better control of crystallisation. One is ΔC, which in practice is set to about a half of the metastable range, the other is the crystal mass m_T or the crystal surface area A, set to values between 15 and 25 wt% expressed as suspension density. As consequence, both factors cannot be taken for influencing the CSD and the mean crystal size.

The still remaining influence factors for modifying product quality are the crystallisation temperature T and the energy input ϵ. With increasing temperature mainly the crystal growth (k_g) will be accelerated whereas the energy input ϵ affects the nucleation rate. Influencing the CSD and the mean crystal size, therefore, remains possible by reducing ϵ and increasing the crystallisation temperature.

A further important factor is the crystal retention time τ, which is related to the linear crystal growth rate G (Eq. (9.4)) and the mean crystal size L using the population balance theory (Eq. (9.5)).

$$G = \frac{dL}{dt} = \frac{1}{\rho} \cdot \frac{\beta}{3\alpha} \cdot \frac{1}{A} \cdot \frac{dm}{dt} \qquad (9.4)$$

$$\bar{L} = 3.67 \cdot G\tau; G \equiv G_k \qquad (9.5)$$

Longer retention times are necessary if larger crystals are desired. But in practice also the opposite can be observed. The reason is the mechanical attrition rate G_a which works against the kinetic growth rate G_k. The effective linear crystal growth rate G_{eff}, therefore, results from $G_k - G_a$. (Eq. (9.6)).

$$G_{eff}(L, \Delta C) = G_k(\Delta C) - G_a(L) \qquad (9.6)$$

$$L = 0; G_a = 0 \Rightarrow G_{eff} = G_k \qquad (9.7)$$

$$L = L_g; G_a = G_k \Rightarrow G_{eff} = 0 \qquad (9.8)$$

Because attrition is intensified with increasing crystal size there always will be a critical size, where the effective crystal growth rate becomes zero (Eq. (9.8)). Therefore, under certain conditions a maximum achievable crystal size exists and longer retention times may lead to smaller crystal sizes.

9.3
Types of Crystallisers

All this is considered in the basic type crystallisers (Figure 9.8) [6]. Crystallisers with longer retention times are able to be operated with less specific energy input, resulting in lower nucleation rates. The impacts between crystals and the impeller pump blades are the most effective source for the nuclei production. These impacts are at least 100-fold more effective than crystal/wall and crystal/crystal impacts.

Types of crystallisers, therefore, differ from each other mainly in design and position (whether operated in clear liquor or in suspension) of the impeller pump (compare Figure 9.5). It can be summarized as follows:

Crystals become bigger with

- decreasing the energy input;
- decreasing the mechanical stresses;
- decreasing the attrition and
- increasing the retention time together with decreasing the attrition.

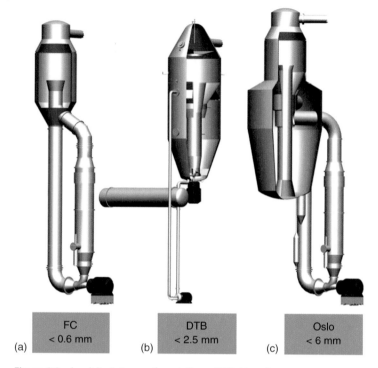

Figure 9.8 (a–c) Basic types of crystallisers (GEA Messo).

Also, with rising temperature the conditions for crystalliation get improved. Besides the specific energy input to a crystalliser system also the energy level takes influence to the nucleation rate. The specific energy input ϵ by impeller pumps can be defined as

$$\frac{dN}{dt} = \frac{\frac{dV}{dt}\rho g H}{\eta}, \quad \epsilon = \frac{\frac{dV}{dt}\rho g H}{\eta}\frac{1}{V_{cryst}} \tag{9.9}$$

Here, ρ denotes the specific density of the suspension or solution; g, the acceleration due to gravity; H, the delivery head; η, the efficiency and V_{cryst}, the filled volume of the crystalliser. In designing the circulation pump, the impeller diameter D and the number of revolutions n of the impeller per unit time can be varied at constant power N:

$$\frac{dN}{dt} \sim n^3 D^5 \text{ or } \epsilon \sim n^3 D^5 \frac{1}{V_{cryst}} \tag{9.10}$$

As a result, the tip speed and, hence, the secondary nucleation rate $B°$ can be altered at a constant rate of dissipated energy ϵ. Industrial-scale experience indicates that

$$V = const \rightarrow \varepsilon = const < \begin{array}{l} n_{large}D_{small} \rightarrow B_{high} \rightarrow \bar{x}_{small} \\ \\ n_{small}D_{large} \rightarrow B_{small} \rightarrow \bar{x}_{large} \end{array} \quad (9.11)$$

'Pumps' with larger diameters (i.e. lower tip speeds and lower secondary nucleation) are used in draft-tube crystallisers for producing coarse crystals.

If mass crystallisation is used as a simple solidification process, only, the production process needs just well-separable crystals. Good separability can be defined as a crystal size of 0.2 mm what allows high-speed impeller pumps. This is realized in FC-type crystallisers (Figure 9.5a). For coarser crystals the energy consumption has to be reduced as it is realized in the so-called draft-tube-baffled crystallisers (Figure 9.5b), which are equipped with draft-tube impeller pumps. These pumps have diameters up to more than 1 m, resulting in much less tip speed compared with external impeller pumps. Additionally, in the case of evaporation crystallisation there is a further strong advantage compared to FC-types. In the (draft-tube-baffled) DTB-crystallisers the heat exchanger which is responsible for most of the pressure loss is taken out of the main circulation. Thus, the central impeller pump consumes only about one-third of energy than those in FC-types, having a strong improving effect on crystal size. The heat exchanger is installed in a second circuit operated with clear liquor, only. Moreover, the operation with clear liquor has an additional improving effect on crystal size. Clear liquor is obtained by withdrawing the liquor from the suspension by sedimentation in a certain part of the crystallizer. Still it contains fines and nuclei. Those fines withdrawn that way extremely reduce the number of crystals in the crystalliser. In DTB-type crystallisers, therefore, mean crystal sizes between 0.5 and 1.5 mm can be achieved.

For production of coarser crystals the principle of circulation a suspension has to be left, as it is realised in the fluidised-bed crystallisers where only clear liquor is circulated (Figure 9.5c) and therefore the nucleation rate drastically reduced. The achievable crystal sizes reach to 5 mm and more.

Among these type classes of crystallisers there are several different crystalliser designs. Figure 9.9 shows a collection of the simplest category.

The stirred-tank crystalliser (1) is selected for vacuum-cooling crystallisation. The horizontal crystalliser (3) is characterised by several series-connected stages in a single shell. Compared to the single-stage, vertical, agitated-tank crystalliser, a horizontal crystalliser of this type has several advantages, mainly savings in investment costs [7]. The draft-tube crystalliser (2) operates with a circulating pump instead of an agitator, and therefore direct control of the process supersaturation is possible. The primary nucleation can therefore be prevented more positively. This type is used for products which grow less fast and have smaller metastable ranges. The forced-circulation crystalliser (4) is comparable in function to the draft-tube crystalliser. The controlled conveying of the suspension is carried out by means of an axial-flow pump through an external heat exchanger. This type of crystalliser can also be used for vacuum-cooling crystallisation (5).

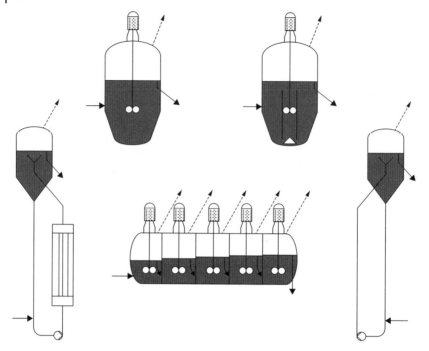

Figure 9.9 FC-type crystalliser (GEA Messo).

Figure 9.10 Crop crystallisation system for a food product (GEA Messo).

Figures 9.10–9.15 give an overview over various FC-type applications. Figure 9.10 is a typical crop crystallisation system for a food product, which needs not to be crystallised as granular product, but must be recovered in a very pure quality and at high yields.

Figure 9.11 Pickling bath regeneration (GEA Messo).

Figure 9.12 Salt crystallisation in Saudi Arabia (GEA Messo).

Figure 9.11 shows the bottom part of a FC-type for the recycling of pickling bath liquor (crystallisation of ferrous sulphate heptahydrate) with a hung recirculation pump in Germany.

Figure 9.12 is a typical configuration for the sodium chloride factory based on dissolution of solar salt with FC-type crystallisation operated by mechanical vapor recompression (MVR) evaporation crystallisation in Saudi Arabia.

Figure 9.13 shows a quintuple-effect evaporation crystallisation plant for the concentration of Yeast Vinasse and recovery of Potassium sulfate crystals in Hungary.

Figure 9.14 is a waste water application of FC-types for the recovery of Sodium sulfate in South Africa, and Figure 9.15 shows the conversion crystallisation of

Figure 9.13 FC in vinasse concentration (GEA Messo).

Figure 9.14 FC for waste water treatment (Na_2SO_4) (GEA Messo).

Potassium sulfate from Chile saltpetre and muriate of potash (MOP) in FC-type crystallisers (inside the building).

With suspension crystallisers, the suspension density is fixed directly by the mass flux. Higher densities (e.g. for achieving a faster desupersaturation rate) can be achieved only if clarified solution is removed from the crystalliser. Such possibility for the removal of clear mother liquor is realized in the group of draft-tube-baffled crystallizers [6, 8] (Figure 9.16). The gentle conveying of the suspension, the removal of the fines and the separate removal of clear liquor, as well as the classifying effect, allow the production coarser crystals. Thus, products such as ammonium sulphate, potassium chloride (see Figures 9.17 and 9.18) or

Figure 9.15 KNO$_3$-plant in Chile (GEA Messo).

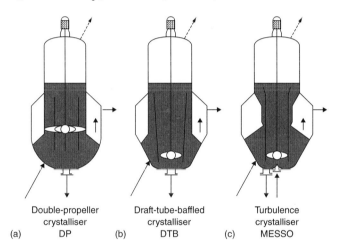

(a) Double-propeller crystalliser DP (b) Draft-tube-baffled crystalliser DTB (c) Turbulence crystalliser MESSO

Figure 9.16 (a–c) DTB-crystallisers (GEA Messo).

urea (Figure 9.19) are produced in this type of crystalliser, with average particle sizes of about 1.5–2.5 mm.

Still coarser products can only be produced in fluidised-bed type crystallisers. Best known is the 'Oslo'-type (Figure 9.20a and b). At present, two forms of this crystalliser type exist. The origin, known by the name 'Krystal', is not without problems in operating behaviour. Incrustations can block the ring gap to the fluidised-bed very soon. The crystalliser must be washed out and reset into operation. The more recent type 'MESSO' [9] was developed especially for the crystallisation of substances prone to form incrustations, and does not have these problems. By reversing the flow in the evaporation section, the superheated solution from the heat exchanger is passed over the cone of the evaporator before evaporation occurs on reaching the solution surface and supersaturation sets in.

Figure 9.17 DTB for urea crystallisation (GEA Messo).

H = 14 m
D = 6 m

Figure 9.18 DTB for KCl at the 'Mines de Potasse d'Alsace' (GEA Messo).

Figure 9.19 DTB for KCl (GEA Messo).

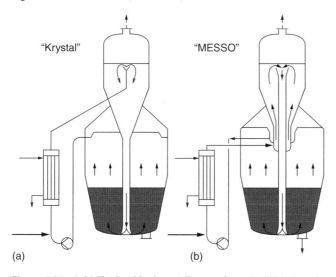

Figure 9.20 (a,b) Fluidised-bed crystallisers; schematic (GEA Messo).

The formation of incrustations on the wall surfaces of the evaporation section is thus avoided, and undisturbed operating times of many weeks can be achieved. Those fluidised-bed crystallisers are required for the production of dust-free products, for example carcinogen products or poisons (Figure 9.21).

9.4 Periphery

As mentioned above, the crystallisation process is not finished with the crystallisation step itself. The created suspension still has to be separated, the crystals have to be dried and packed. The vapours are to be condensed and the non-condensable

Figure 9.21 Fluidised-bed crystalliser (GEA Messo).

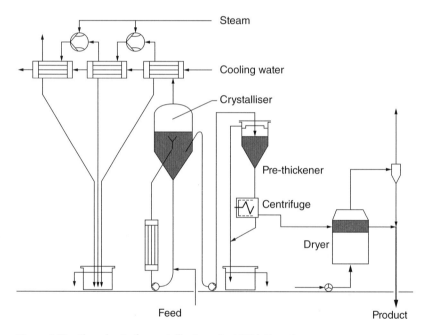

Figure 9.22 Flow sheet of a crystallisation plant (GEA Messo).

gases taken off by vacuum pumps. Figure 9.22 shows a simplified flow sheet of such a complete crystallisation plant operated for vacuum evaporation crystallisation.

Instead of the FC-type crystalliser any other type of crystalliser can be installed. Also multiple-effect units can be taken. Or, instead of using steam for heating it could be decided for mechanical or thermal vapour re-compression. In this example the vapours from the (last) crystalliser are condensed in a

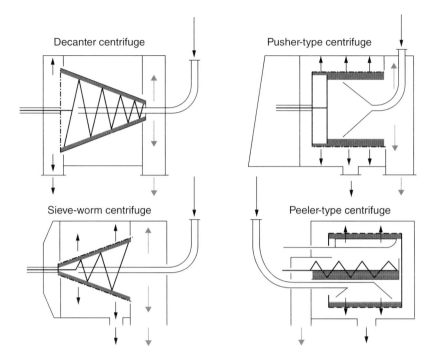

Figure 9.23 Centrifuges (GEA Messo).

surface condenser. Also a mixing condenser could be chosen. The suspension can be withdrawn by overflow, as shown. The suspension still is too low concentrated regarding the best operating conditions for the centrifuge. The suspension densities are between 15 and 25 wt% whereas centrifuges are operated best at 50–60 wt%. Therefore, the suspension at first is pre-concentrated in pre-thickeners or hydrocyclones. The clear liquor overflow is recycled. Part of it always has to be taken as purge liquor for bringing out the system impurities. The underflow is guided to the centrifuge for separation.

Depending on the CSD there is the choice between approximately four types of centrifuges, the decanting type and peeler type for the finer products, the sieve-worm type and the pusher type for normal products (Figure 9.23). The final drying of the product in most of the cases is proceeded in draft-tube or vibrating fluidised-bed dryers (Figures 9.24 and 9.25).

9.5
Process Particularities

Besides the already mentioned criteria the design of the crystallisation processes is influenced by several additional factors. These are explained in the following sections.

Figure 9.24 Vibrating fluidised-bed dryer (NaCl) (GEA Messo).

Figure 9.25 Vibrating fluidised-bed dryer (CaCl$_2$) (GEA Messo).

9.5.1
Surface-Cooling Crystallisation

The surface-cooling produces supersaturations directly on the heat exchanger surface. This supersaturation is the highest in the entire crystalliser. Incrustations of the cooling surface and a limitation of the plant operation are the normal consequences. That can be accepted for discontinuous operation, because with the next batch all the incrustations get dissolved again. For continuous processes the surface cooling is only decided for, if too low operating pressures make vacuum cooling crystallisation uneconomically. One reason could be a high boiling point elevation. In that cases especial large heat exchanger surface are chosen.

9.5.2
Vacuum-Cooling Crystallisation

Vacuum-cooling crystallisation is the preferred cooling crystallisation method under continuous operation. Because cooling is generated by adiabatic expansion of the solvent no cooling surfaces can be incrusted. Vacuum-cooling becomes uneconomical only if cooling has to be effected at very low temperatures.

9.5.3
Evaporation Crystallisation

The evaporation crystallisation is a vacuum process as well as the vacuum-cooling crystallisation. Differing from the vacuum-cooling this process method is independently from the concentration and temperature of the feed solution. External heat can be added to the system and the concentration of mother liquor can be adjusted independently. Like vacuum-cooling crystallisation, there are no special incrustation problems in evaporation crystallization if boiling on the heater surface is prevented. Some difficulties may arise for inverse soluble substances, like the hardeners. In those cases it has to be taken care as explained for the surface-cooling crystallisation. High velocities of the suspension flow and a high suspension density can be helpful by erosion and faster desupersaturation behaviour. For a better process economy the units are constructed as multiple-effect evaporation plants.

9.6
Example – Crystallisation of Sodium Chloride

Nowadays, there is the possibility to combine solar evaporation with modern salt crystallisation for improving salt quality. Figures 9.26 and 9.27 show the flow sheet of a sodium chloride crystallisation unit, which is energy saving-wise operated with concentrated liquor from a solar pond.

Besides normal FC-types there is installed one fluidised-bed crystalliser for producing a coarse quality as well. The plant is operated as multiple-effect evaporation unit in sequence. The Oslo-type crystalliser is set as the first effect ensuring best conditions for crystal growth at the highest temperature. The plant is fed with concentrated liquor from the solar pond into tank B, from where the Oslo-type crystallizer gets served. This tank is operated undersaturated ensuring crystal-freeness in the feed to the Oslo-type crystalliser, otherwise the production of coarse crystals is getting endangered. Solution not used for the Oslo-type overflows into tank C, from where the FC-type crystallisers get fed. The slurry withdrawal from the FC-type crystallisers is taken from stage to stage and from the last effect to tank A.

From there the separation station is served which consists of a hydrocyclone, a washing-thickener and a centrifuge. In the washing thickener, the highly impure mother liquor is exchanged against the fresh feed liquor, before the suspension

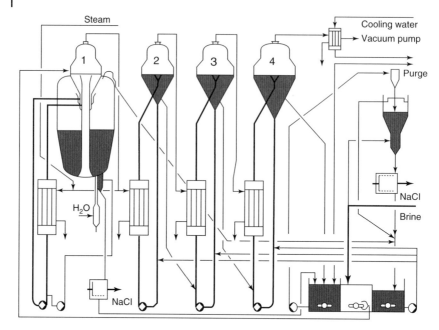

Figure 9.26 Concept for a salt factory (GEA Messo).

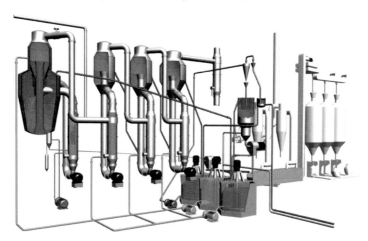

Figure 9.27 Concept for a salt factory (3D) (GEA Messo).

is given to the centrifuge, finally. On the centrifuge the salt is washed with water, then dried and packed. The separation of the grain-size product is effected by a special pusher-type centrifuge, ensuring a soft treatment of this sensitive product.

A plant of this type was erected for the treatment of 60 t h^{-1} concentrated liquor from a solar pond on the coast of the Mediterranean Sea. The plant is producing 2.5 t h^{-1} grain-size salt of ≥2 mm average crystal size and 6.5 t h^{-1} of normal vacuum salt. Per hour it consumes 11 t of steam and evaporates 34 t of water. By harvesting the concentrated liquor in special tanks over the summer period, the plant is able to be operated the entire year.

References

1. Bamforth, A.W. (1965) *Industrial Crystallization*, Leonard Hill, London.
2. Matz, G. (1969) *Kristallisation*, Springer-Verlag, Berlin, Heidelberg, New York.
3. Mullin, J.W. (1993) *Crystallization*, 3rd edn, Butterworth-Heinemann, Oxford.
4. Nyvlt, J. (1971) *Industrial Crystallization from Solutions*, Butterworth, London.
5. Mersmann, A. and Kind, M. (1985) *Chemie-Ingenieur-Technik*, **3**, 190.
6. Wöhlk, W. and Hofmann, G. (1985) *Chemie-Ingenieur-Technik*, **4**, 322.
7. Messing, T. and Hofmann, G. (1980) *Chemie-Ingenieur-Technik*, **11**, 870.
8. Kasai, Tatusi, Yokohama (1966), Klassifizienden Kristallisator, Patent DE-PS 1 519 915, Tsukishima Kikai Co., Ltd., Tokyo.
9. Hofmann, G. (1983) Ein Oslo-Kristallisator für lange Reisezeiten, Lecture on VDI/GVC-Meeting "Kristallisation", Deggendorf, Bavaria, 3/1983.

10
Why Evaporation under Vacuum?

Gregor Klinke

Summary

Evaporation plants should be operated at high pressure, if possible, as in this way the vapours can be used to the best possible extent. But there are many other reasons why evaporation plants should be operated under vacuum.

Based on different criteria we will show the cases in which vacuum operation is advisable or even necessary.

10.1
Introduction

Evaporation is the separation technology to separate a solvent – in most of the cases water – from a solution, emulsion or suspension by means of boiling, while the dissolved or suspended substance has no or only a slight steam pressure.

It is the objective of this separation technology to recover the solvent as pure as possible, for example, in the production of drinking water from sea water.

In other cases, the concentration of the dissolved, emulsified or suspended substance is given priority. For example, for the production of condensed milk, it would be desirable to achieve a higher dry substance contents in the product.

Another objective of evaporation is the oversaturation of the solution until precipitation of crystals. For the production of kitchen salt, for example, sodium chloride solution is evaporated in order to product salt crystals.

Frequently both objectives, that is, concentration of the solution and production of a possibly pure solvent are expected from an evaporation plant.

10.2
Thermodynamics of Evaporation

Evaporation is a thermal separation process, that is, heat has to be supplied to the liquid. This is performed in two process steps and in different apparatuses.

10 Why Evaporation under Vacuum?

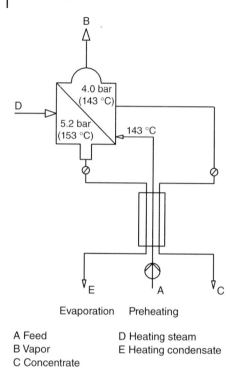

A Feed
B Vapor
C Concentrate
D Heating steam
E Heating condensate

Figure 10.1 Example for ideal pressure evaporation.

First, heat is supplied to the preheater in order to heat the product to boiling temperature. In the second step, heat is supplied in the evaporator in order to evaporate the solvent.

The steam separated from a solution is also referred to as vapour. The temperature of the vapour corresponds to the saturated steam pressure of the solvent and the boiling point elevation caused by the dissolved substance.

In most of the cases, water steam is used as heating medium. The description below is based on the fact that water steam is the heating medium.

For the heat transfer in the evaporator, the condensation temperature of the heating steam has to be higher than the boiling temperature of the product.

The heating steam condensate and product concentrate flows, which are discharged from the plant are available for product preheating. In the ideal case as shown in Figures 10.1 and 10.2, both flows are sufficient to preheat the product. In this case, no additional energy is required for the preheating.

The ideal case also illustrates that during evaporation the vapour energy equals the heating steam energy; however, the entropy increases at the same time. The pressure and temperature level of the vapour is lower. The ideal evaporation does not consume any energy, heating steam energy is required and vapour energy is supplied. The cost efficiency of an evaporation plant depends on how you can

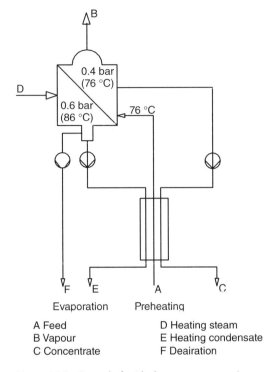

Figure 10.2 Example for ideal vacuum evaporation.

use the vapour energy. The possibility of use depends on whether the vapour is supplied under pressure or vacuum.

10.3
Pressure/Vacuum Evaporation Comparison

At the first glance it seems that pressure evaporation should be preferred at any rate since vapour can be used best under pressure. However, for the operation of an evaporation plant different criteria with economical and substance-specific aspects have to be considered.

- Vapour utilisation
- Design of the apparatuses
- Machine equipment
- Corrosion
- Insulation
- Safety aspects
- Product properties and
- Boiling range.

Figures 10.1 and 10.2 show examples of a pressure and vacuum evaporation.

The comparison of the two processes results in different advantages of the individual systems, depending on the criterion.

10.3.1
Vapour Utilisation

When using the vapour, you always will aim at a sufficiently high pressure level, because with decreasing pressure the condensation temperature of the vapour will decrease, and therefore, the chance to use the vapour for heating purposes will be also reduced. In case of a very low vacuum it will be just possible to condense the vapour via cooling water, the vapour energy will be dissipated to the environment, which means that it is lost.

A high vapour pressure requires a correspondingly high heating steam pressure. Moreover, high-pressure heating steam is very expensive.

If low-price, low-pressure exhaust steam can be used, you will inevitably have a vacuum evaporation.

10.3.2
Design of the Apparatuses

In case of pressure evaporation, the apparatuses of course have to be designed according to the overpressure. This causes increased wall thickness of the apparatuses. Moderate overpressures must not inevitably result in higher plant expenses, because a reduced design is possible for pressure evaporators in partial areas, and vacuum units require corresponding reinforcements. Depending on the pressure graduation and on the size of the apparatus, either pressure operation or vacuum operation is more advantageous. At any rate, pressure evaporators are plants, which are subject to monitoring.

10.3.3
Machine Equipment

The comparison of Figures 10.1 and 10.2 shows that a vacuum plant requires a more sophisticated machine equipment than a pressure plant. On the heating and product side, vacuum pumps are required and liquid pumps are required to discharge the heating steam condensate and the product concentrate. Theoretically, fresh product can be sucked in via vacuum; depending on the type of evaporator another pump will be required for the feed.

For the pressure evaporation, the feed pump only will be required. The discharge of product and heating steam condensate is realized by the system pressure. Simple condensate traps are used to maintain the pressure for the condensate, on the product side, usually a level-control loop will be required.

As a matter of course, pressure apparatuses have to be equipped with safety equipment.

10.3.4
Corrosion

The chemical corrosion attack increases exponentially with high temperatures. Also for this reason, there are limits for the evaporation temperatures.

10.3.5
Insulation

In order to keep loss of heat as low as possible, evaporators have to be insulated at high temperatures. In many cases, evaporation plants in buildings, which operate at maximum 60 °C are not insulated.

10.3.6
Safety Aspects

The advantage of a vacuum-operated plant is the fact that the evaporation will stop immediately if the vacuum is interrupted. In case of faults, a safe operation can be established immediately.

This is not the case in pressure-operated plants. A pressure-less operating mode can only be achieved if heat in form of flash vapour and product is discharged and this always takes a certain time.

Blowing off of the safety valves can result in predictable faults. These faults can be avoided by setting up an emergency condensation.

The pressure evaporation will represent a problem in case of unpredictable faults, such as leaks for which a directed discharge is not possible. If hazardous substances are processed, a pressure evaporation can therefore not be used.

10.3.7
Product Properties

The properties of the product to be evaporated precede all previous criteria. The maximum possible concentration of a product without modification of the product and without coating on the exchange surfaces largely depends on the operating temperature and on the pressure.

Usually, a product can be concentrated better at higher temperatures, because the viscosity is reduced with increasing temperature. Figure 10.3 shows the influence of the temperature on the viscosity of water.

Watery solutions or suspensions show the same trend. Therefore, always a high operating temperature has to be aimed at.

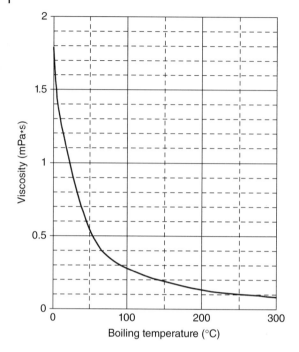

Figure 10.3 Viscosity of water.

However, for many substances these possibilities are limited because they change colour with a certain temperature or they decompose or coagulate or coat the exchange surfaces.

For this reason, in the food industry only vacuum evaporation plants are used.

10.3.8
Boiling Range

The above-mentioned criteria only refer to watery solutions, but those with a high boiling point also have to be mentioned. These are, for example, saturated hydrocarbons with more than six C-atoms. For an atmospheric evaporation heating temperatures of up to 300 °C are required.

Low boiling temperature and thus, low heating temperatures can be also achieved with vacuum.

For diphenyl ($C_{12}H_{10}$), for example, the boiling temperature can be reduced from 255.9 to 150.7 °C by reducing the pressure from 1000 to 100 mbar.

A vacuum operation is absolutely required for substances with a high boiling point.

10.4 Possibility of Vapour Utilization

10.4.1 External Utilization

The vapour of evaporation can be fed into a steam system or can be supplied to another consumer. However, the capacity of the evaporator always depends on the consumed vapour quantity.

In most of the cases, the vapour is used within an evaporation plant, which is designed according to the requested capacity. There are generally two possibilities: the multi-stage evaporation and the vapour recompression.

10.4.2 Multi-Stage Evaporation

The vapour can be used by being supplied as heating steam to a second evaporator stage, which works at a lower pressure. The vapour from this can be supplied to a third stage, and so on. In this way, in case of n stages in the ideal case n kg of vapour can be evaporated with 1 kg of heating steam.

The required pressure graduation is generally carried out in the vacuum range, as far as the vapour of the last stage still can be condensed with cooling water.

The heat transfer surfaces of the individual stages have of course to be dimensioned accordingly. With increasing number of stages the temperature difference per stage will get lower and lower. Their heating surface then has to be dimensioned larger. In a first approximation, you can say that with increasing number of stages the total heating surface of a plant will increase proportionally, and the investment costs will increase considerably. Therefore, the design will aim at an optimum between investment costs and operating expenses.

The plant shown in Figures 10.4 and 10.5 is a four-stage plant; theoretically it evaporates 4 kg of vapour with 1 kg of heating steam; this implies a specific consumption of thermal energy of 25%.

10.4.3 Mechanical Vapour Recompression

For the multi-stage evaporation, the vapour energy is used via a pressure reduction from stage to stage. The vapour energy can also be used if the vapour is compressed to the higher pressure in the heating chamber. With increasing compression also the condensation temperature of the vapour increases so that it can be used for the heating of the same evaporator stage. Picture of a plant in Figure 10.6.

The evaporation with mechanical vapour recompression corresponds to the ideal evaporation; the vapour energy is completely used since no pressure graduation is required as it is the case for the multi-stage evaporation. Apart from the starting steam only electric energy is required to drive the compressor.

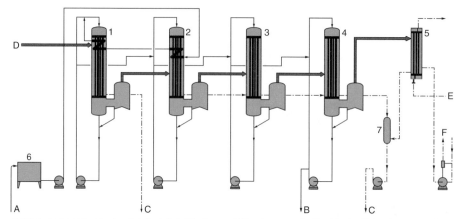

1,2,3,4　Evaporation body, 1st, 2nd, 3rd, 4th stage,　5 Condenser,　6　Feed tank,　7 Condensate tank
A　Fresh product,　B Concentrate,　C　Condensate,　D　Fresh steam,　E　Cooling water,　F　Ventilation

Figure 10.4　Vacuum falling-film evaporation plant.

Figure 10.5　Four-stage falling-film evaporation plant.

Figure 10.6 Evaporation plant with mechanical vapour recompressor.

The energy consumption required for compression also depends on the pressure. Figure 10.7 shows the 'Specific compression work' in kilowatt hour per degree of saturation temperature increase and per ton of water steam.

At a working pressure of approximately 200 mbar, for example, theoretically a work of 2.0 kWh degree^{-1}*ton will be required. With a compression efficiency of 80% an electrical work of 2.5 kWh degree^{-1}*ton would have to be supplied.

When working with a temperature increase of 6 degrees, with a corresponding design of the evaporator an electrical energy requirement of 15 kWh ton^{-1} of water evaporation is required.

For a comparison with thermal energy, this figure has to be corrected with the power station efficiency of, for example, 38% resulting in a water evaporation of *39 kWh ton^{-1}*. This figure can be compared to energy consumption in the form of steam of a water evaporation of *625 kWh ton^{-1}* in case of a single-stage evaporation without the use of vapour. This corresponds to a consumption of thermal energy of only *6.24%*.

In a multi-stage plant such values cannot be achieved. However, this important advantage in terms of energy costs has to be paid via correspondingly large evaporator surfaces – that is, the investment costs are higher. Moreover, the mechanical vapour recompression can only be used for products with no or only slight boiling

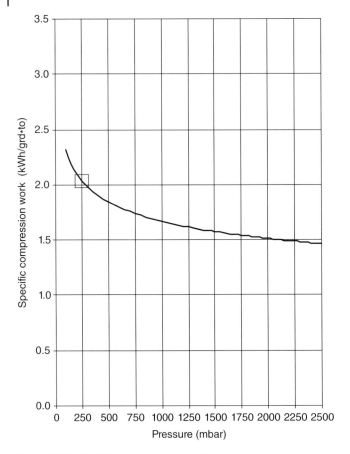

Figure 10.7 Specific compression work for water steam.

point elevations, such as milk or whey. As shown in Figure 10.7, the 'specific compression work' is reduced with increasing working pressure. The energy costs with mechanical vapour recompression could be further reduced via operation in the pressure range. However, most of the evaporators are operated under vacuum for the reasons, which are explained above.

Further Reading

Baehr, H.D. (1962) *Thermodynamik*, Springer-Verlag, Berlin, Göttingen, Heidelberg.

Bosnjakovic, F. (1972) *Technische Thermodynamik II.Teil*, Verlag Theodor Steinkopff, Dresden.

GEA WIEGAND GmbH, Ettlingen, Evaporation Technology (1997) *firm's brochure P03E 052004 VDI-Wärmeatlas*, 8th edn, Springer-Verlag.

Rant, Z. (1959) *Verdampfen in Theorie und Praxis*, Verlag von Theodor Steinkopff, Dresden und Leipzig.

Spalding, D.B., Traustel, S., and Cole, E.H. (1965) *Grundlagen der technischen Thermodynamik*, Friedrich Vieweg & Sohn, Braunschweig.

11
Evaporators for Coarse Vacuum
Gregor Klinke

Summary

For evaporation under coarse vacuum a variety of evaporator types is available. They represent a historical development, and each type has its specific field of application. Some typical evaporator types are described here.

11.1
Introduction

Working pressures between 80 and 800 mbar abs characterise the coarse vacuum. Water temperatures between 42 and 93 °C correspond to these pressures. Most of the watery solutions are evaporated in this range. The evaporator types which are used here are hardly different from those used in the pressure range.

Exhaust steam processes with liquids, however, are mainly operated in the fine vacuum range down to pressures of 1 mbar. The construction of the evaporators which are used here is very different from that of evaporators working under coarse vacuum.

11.2
Criteria for the Selection of an Evaporator

For the selection of an evaporator different substance-specific and economical criteria have to be considered.

11.2.1
Suitability for the Product

This aspect has first priority. Evaporators for sensible products, such as food and pharmaceutical products are an example for this. Microbe growth has to be avoided as much as possible.

Vacuum Technology in the Chemical Industry, First Edition. Edited by Wolfgang Jorisch.
© 2015 Wiley-VCH Verlag GmbH & Co. KGaA. Published 2015 by Wiley-VCH Verlag GmbH & Co. KGaA.

The plants require a sanitary design, that is there must be no centre of infection, the product must not form residuals in the plant, the plant has to be rinsed well in all points.

In order to reduce the microbe growth, the residence time of the product in the plant should be defined, a back-mixing for example, is very detrimental. This can be ensured by a single pass of the product.

Short residence times and small filling volumes are requested for the same reason.

A foaming in the evaporation plant can endanger the complete process step. The selection of the right evaporator is the solution for this problem, because antifoaming agents are not always efficient or are not allowed.

11.2.2
Cleaning

Contamination and coating on the heat-transfer surfaces are undesirable phenomena in the evaporation plant. The heat transfer is reduced, residues result in an increased microbe growth.

Coating can be avoided by a good rinsing of the transfer surfaces and by small temperature differences.

However, a plant has to be cleaned at any rate. Evaporators for the pharmaceutical and food industry have to be cleaned every day for bacteriological reasons.

The cleaning time reduces the production time and should therefore be short. In the best case, you can clean the plant with hot condensate or with a watery caustic soda and/or nitric acid, without opening the plant.

If the plant has to be opened for a mechanical cleaning, for example with high-pressure nozzles, the surfaces have to be easily accessible. Tube evaporators show evident advantages in this respect.

11.2.3
Quality of Heat Transfer

If you decide in favour of a certain temperature difference for heat transfer, the k value will determine the size of the transfer surface and thus, the price of the plant. The differences for the individual evaporator types are enormous. This knowledge frequently is the manufacturer's secret.

11.2.4
Required Space

The space requirements of an evaporation plant can be important. The investment for an evaporator is therefore connected with the space requirement, independent of whether the plant shall be installed in a building or in the open air.

On principle, there are two variants:

- Plate evaporators can be installed on one level, and therefore the required floor space is large.
- Falling-film evaporators do not require a large floor space, they require a certain height and therefore, several operating platforms are necessary.

11.2.5
Cost Efficiency

The operation of an evaporation plant will be cost efficient

- if it delivers the product in the requested quality,
- if it is available for a maximum period of time and
- if the operating expenses are as low as possible.

Sometimes, the price for the plant is only of secondary importance.

11.3
Evaporator Types

11.3.1
Agitator Evaporator

The origin of these evaporators is evident. It is a large-scale cooking pot with installed agitator. Process engineering systems frequently are equipped with these heated agitator tanks. In most cases, they are used in discontinuous operation as reactors, mixers, crystallizers and finally as evaporators. With their large filling volume they are predestinated, for example as waste water collection tank, which can be operated as evaporator if required to concentrate the product. With the installed agitator, the product can be concentrated to a considerable viscosity. Picture of a plant in Figure 11.1.

Fruit juices and marmalades are concentrated in this way. Thanks to the simple design they are sanitary and can be cleaned easily.

11.3.2
Natural Circulation Evaporator

The natural circulation evaporator is the classic evaporator type. In order to achieve larger heating surfaces, the agitator evaporator was soon replaced by the tube evaporator. The vertical arrangement with relatively short tubes results in a natural circulation of the product in the system separator/heater.

Figure 11.2 illustrates the separator (2) in which the liquid is in boiling state. If it reaches the lower heating body (1) via the circulation tube (3), it will not boil due to the additional pressure of the liquid. Compared to the current pressure, it

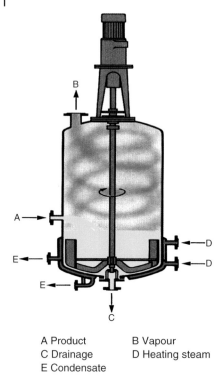

A Product B Vapour
C Drainage D Heating steam
E Condensate

Figure 11.1 Agitator evaporator.

is sub-cooled and has to be heated in the lower heating body part first. This is done at laminar speeds; therefore, the heat-transfer values are bad.

After this heating section and, above all, with decreasing pressure, the liquid starts boiling. The resulting bubbles mix the liquid. The heat transfer will improve.

In a third step, the steam volume is larger than the liquid volume. The density of the mixture is decreasing. The thermosiphon effect is initiated by the density difference in the heating body compared to the circulation tube. At the same time, the steam moves the liquid through the heating tube at a high speed. The heat transfer is getting extremely good.

Figure 11.3 is a sketch of the process in one single rising tube.

The preheating section reaches temperature t^* at which the product starts boiling. Above this point, you have a phase mixture.

If you install an upright stand pipe to the boiling pipe, you will measure a virtual liquid level Hsch which reaches above the point of t^*. The length of Hsch essentially reflects the length of the preheating zone, the zone of an insufficient heat transfer. The remaining tube length has a very good heat transfer. Hsch can therefore be used as quality index of the heat transfer.

By throttling, the circulating flow Hsch can be reduced and the overall heat transfer can be improved. However, the speed in the liquid phases will still be

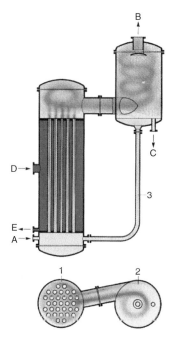

A Product   B Vapour
C Concentrate D Heating steam
E Condensate
1 Heating body 2 Separator
3 Circulation tube

Figure 11.2 Natural circulation evaporator.

reduced by this, and the evaporation rate will increase with every circulation. However, also the risk of coating is increasing.

The different phases in the rising tube and the bulged temperature profile as shown in Figure 11.3 makes the calculation of the natural circulation evaporator difficult. The temperature difference over the tube length is very different, and it is rather difficult to determine an average value.

Nevertheless, natural circulation evaporators are commonly used.

11.3.3
Climbing-Film Evaporator

The climbing-film evaporator is the further development of the natural circulation evaporator. Before entering the evaporator, the product is preheated to boiling temperature or to a temperature slightly above boiling temperature, which is even better. Evaporation takes place immediately on the product inlet. In this way, the complete tube grid is available for the two-phase heat transition with high k values.

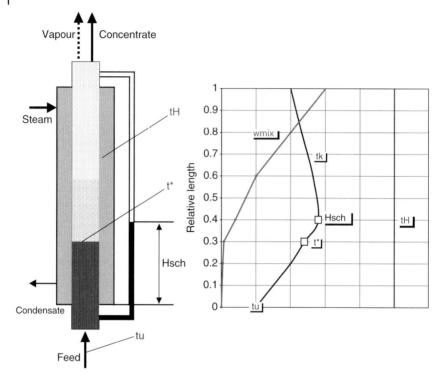

Figure 11.3 Boiling in rising tube.

The heat transition is improved with increasing steam speed in the tube, in practice, tubes up to a length of 8 m are applied in order to take advantage of this effect.

The climbing-film evaporator is a high-performance evaporator.

The fact that the vapour has to push and accelerate the liquid drops, however, results in a very high dynamic pressure loss, and this again implies that a large temperature difference has to be provided for. For this reason, climbing-film evaporators are not suited for temperature-sensitive products. For the first time, however, this is an evaporator that can be run in single-pass product flow (Figure 11.4).

11.3.4
Falling-Film Evaporator

The disadvantage of the high pressure drop in the climbing film will be removed to a large extent if operating with a falling film. In the falling-film evaporator, the product descents as film on the inner tube wall in the gravity field of the earth and is accelerated only moderately by the vapour (Figures 11.5 and 11.6).

The pressure drop is lower and evaporation tubes can be designed up to lengths of 20 m.

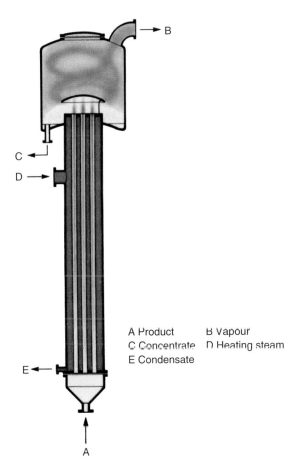

Figure 11.4 Climbing-film evaporator.

A Product   B Vapour
C Concentrate   D Heating steam
E Condensate

The advantage of the falling-film evaporator over the climbing-film evaporator is that of a moderate pressure drop. Thanks to the low pressure drop, it is very suitable for a multi-effect evaporator or for plants with vapour recompression.

A falling-film evaporator can be operated at low temperature differences, therefore, it is very suitable for temperature-sensitive products; moreover, it can be operated with minimum filling volume and in single pass. For these reasons, the falling-film evaporator has been used with great success in the food industry for the evaporation of milk and whey.

The falling-film evaporator requires additional equipment: the liquid distribution (Figure 11.7).

In a climbing film, the liquid is evenly distributed over the tubes of a heating grid by means of the static pressure. In case of a falling film, however, this equipment is required. In particular in case of small liquid quantities, the distribution must be very uniform. Figure 11.7 shows the distribution via a perforated bowl. The

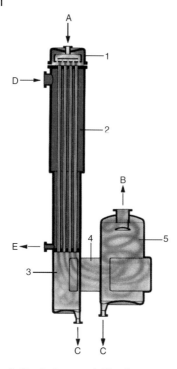

A	Product	1	Head
B	Vapour	2	Heating body
C	Concentrate	3	Heating body base
D	Heating steam	4	Mixture channel
E	Condensate	5	Separator

Figure 11.5 Falling-film evaporator.

boring provides for a static liquid pressure, which makes possible the distribution over the complete surface of the tube sheet.

At this point, it has to be stated that the usually uniform pressure drop of a falling-film evaporator increases with the vapour speed, in particular at medium and low vacuum. In such cases, shorter tubes with larger diameter have to be used. In the fine vacuum range, the use of tube evaporators is avoided, and constructions are used in which the vapour path between evaporator and condensation surface is extremely short.

11.3.5
Forced-Circulation Evaporator

The application of natural circulation, climbing-film and falling-film evaporators will be limited if the product is very viscous or if it strongly tends to incrustations. In this case, a forced-circulation evaporator can be used. In this evaporator type,

Figure 11.6 Four-effect falling-film evaporation plant.

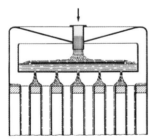

Figure 11.7 Perforated bowl – distribution.

evaporation does not take place in the tube, but the liquid is heated. The actual evaporation takes place by flashing in the separator.

The heat transition which in the natural circulation evaporator is rather low in the liquid phase, is improved by means of a circulation pump. The pump presses up the product through the heating grid with tube speeds of up to $3\,\mathrm{m\,s^{-1}}$. This pump can be arranged vertically, or as shown in Figure 11.8, horizontally.

The high speeds achieve similar good k values as in the falling-film evaporator. Coatings are rinsed off from the tube wall at the high-flow speeds.

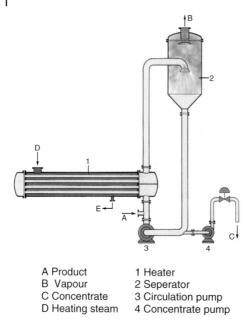

A Product
B Vapour
C Concentrate
D Heating steam

1 Heater
2 Seperator
3 Circulation pump
4 Concentrate pump

Figure 11.8 Forced-circulation evaporator.

In order to avoid coatings, the evaporation in the heating body is suppressed by placing the separator on a higher position as shown in Figure 11.8, or in other cases by means of a throttling orifice downstream of the heating body.

With forced-circulation evaporators, products can be concentrated until they can be pumped. It has to be made sure that the supply of the product to the circulation pump is good.

11.3.6
Fluidised-Bed Evaporator

There is a variety of different incrustations on the heat-transfer surfaces. Some coatings are viscous, others are brittle. Lime and gypsum are such brittle coatings; frequently, they are contained in small quantities in products, but the coatings are rather unpleasant.

If the viscosity of the liquid is not excessive, lime and gypsum coatings can be avoided with a fluidised-bed system. For this purpose, the 'forced-circulation'-type evaporator with vertically arranged heating grid is used (Figure 11.9).

Below the heating grid there is a fluidized bed, which is swirled by the circulating flow of the circulation pump. The swirling bodies reach the heating tubes and clean the inner surface of the tubes with their edges. The chamber above the heating

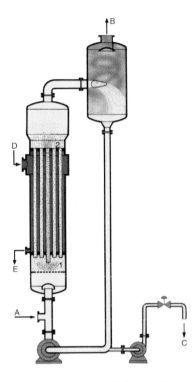

A Product
B Vapour
C Concentrate
D Heating steam
E Condensate

1 Inlet chamber
2 Outlet chamber

Figure 11.9 Fluidized-bed evaporator.

body is expanded, the swirling bodies deposit and return to the lower part through a return tube.

11.3.7
Plate Evaporator

Plate units were first known as liquid/liquid heat exchangers.

Due to the corrugation of the transfer surfaces, a high-flow turbulence and high heat-transfer coefficients could be achieved. Moreover, with a plate heat-transfer unit it was possible to adjust the surface by changing the number of plates. Arrangement of a plant in Figure 11.10.

In the meantime, these advantages are also used for evaporation. The achieved k values are similar as those achieved in a tube climbing-film evaporator. A separate preheater is not required since the heat transfer in the liquid phase is already good.

11 Evaporators for Coarse Vacuum

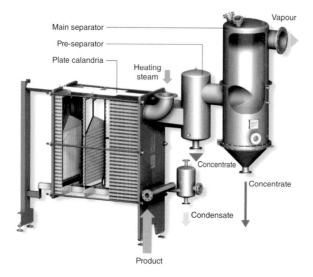

Figure 11.10 Section of a plate evaporator.

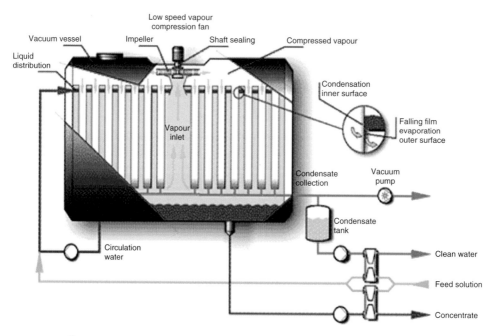

Figure 11.11 HADWACO evaporator system.

A throttling of the evaporation is not recommended since the plates only can take up a limited pressure difference between heating and product side. The plates which support each other are imprinted. Thus, they can resist to higher pressure differences, but the supporting points are always a starting point for the formation of coatings.

The distribution of the liquid over the plates is very important. The complete transfer surface has to be flushed well in order to avoid incrustations. Plate evaporators can be chemically cleaned, but a mechanical cleaning with high-pressure ejector is very difficult. For this reason, plates are only used for evaporation for pure liquids, such as sugar juice.

Plates are also operated as falling-film evaporator.

A development is very interesting, in which welded heating boxes are inserted into tanks. The boxes are wetted with the liquid to be evaporated from the outside.

HADWACO manufactures the boxes even as plastic foils, as a development of this idea, however, they are very limited with regard to temperature and pressure difference. Under vacuum, however, and in connection with a mechanical vapour recompressor with a low pressing (Figure 11.11) the use of plastic bags is possible. An astonishing solution!

Further Reading

GEA WIEGAND GmbH (1998) Ettlingen, Evaporation Technology firm's brochure P03E 052004.

HADWACO LTD OY (1995) Helsinki, Finnland, firm's brochure.

12
Basics of Drying Technology
Jürgen Oess

12.1
Basics of Solids–Liquid Separation Technology

During the synthesis of many chemical and pharmaceutical products, intermediate or final, solids particles are suspended in liquids and have to be separated from these slurries at some stage.

Typically mechanical separators like filter centrifuges, vacuum or pressure filters, decanters, filter presses or simple static thickeners are used to isolate the solids from the liquid.

The optimum type of mechanical separator that is selected for individual processes mainly depends on the properties of the solids, the required capacity and the mode of operation.

However it is in most cases not possible to achieve the required product quality only with one 'Mechanical' Solids–Liquid Separation step, as it not possible to remove all the liquid just by mechanical forces. Depending on the product and the equipment used, the moisture level of discharged products from centrifuges or filters is in a range of 5–80% by weight (Figure 12.1).

This remaining amount of liquid has to be removed in a second, the 'Thermal' Solids–Liquid Separation step which is commonly known as 'Drying'.

12.2
Basics of Drying Technology

Drying is a process to remove liquid from solids using a thermal method.

Contrary to the above-mentioned mechanical Solids–Liquid Separation modes like pressing, filtration or sedimentation where both materials maintain their physical conditions, occurs during the drying process a phase transition of the liquid phase into a gaseous stage.

By applying heat to the product under certain conditions, the liquid is transferred into the gas phase and then removed from the solids to isolate the materials from each other (Figure 12.2).

Vacuum Technology in the Chemical Industry, First Edition. Edited by Wolfgang Jorisch.
© 2015 Wiley-VCH Verlag GmbH & Co. KGaA. Published 2015 by Wiley-VCH Verlag GmbH & Co. KGaA.

12 Basics of Drying Technology

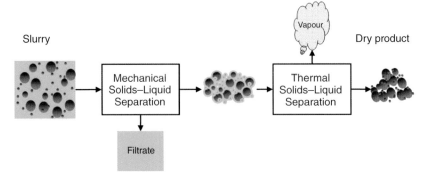

Figure 12.1 Solid–Liquid Separation chain.[1]

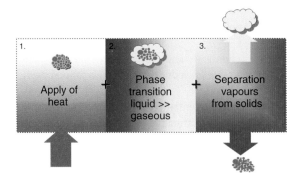

Figure 12.2 Process sequences during drying.

The three process sequences, 'Apply of Heat', 'Phase Transition' and 'Removal of Vapours' are taking place at the same time and at the same location during the drying process.

Depending on the way to apply heat to the product the drying is differentiated in several different modes. The main drying modes are the 'Convection Drying' and the 'Contact Drying' but also 'Radiation Drying' is becoming more popular.

12.2.1
Convection Drying

During Convection Drying, a hot carrier gas is introduced; this either flows over or through the product. Heat is transferred from the gas to the product. The liquid volatilising from the solids is carried away by the gas that is cooling down during this process (Figure 12.3).

Convection drying is typically carried out under atmospheric pressure with air or nitrogen as carrier gas in flash dryer, fluid bed dryer, tray dryer or belt dryer.

1) Product Presentation Heinkel Process Technology GmbH. Ferdinand Porschestr. 8, D-74354 Besigheim.

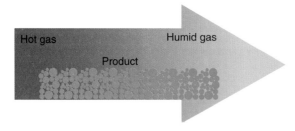

Figure 12.3 Principles of convection drying.

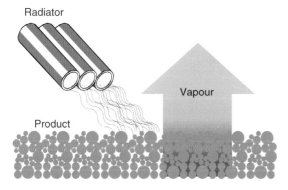

Figure 12.4 Principles of radiation drying.

Convection drying is probably the most common way to dry non-hazardous water-wet products in industrial production.

12.2.2
Radiation Drying

The heat transfer by radiation is possible in several ways. Two well-known methods are the sun-drying and heating up of products inside ovens (kitchen stove) (Figure 12.4).

In the last couple of years, equipment was developed to treat products by irradiation with special wavelengths in a controlled way. Examples are microwave or infrared applications that are also used for the drying of special bulk materials

12.2.3
Contact Drying

12.2.3.1 Heat Transfer during Contact Drying
During contact drying the heat is transferred from heated surfaces to the product.

Depending on the required heating temperature, the walls are either heated by liquid heat transfer media like water or thermal oils, saturated steam or even electrically (Figure 12.5).

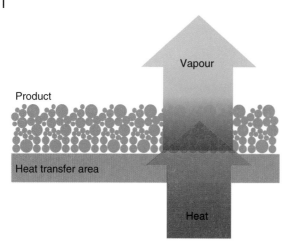

Figure 12.5 Principles of contact drying.

The effective heat flow depends on the available heat transfer area (A), the temperature gradient (ΔT) between the heating medium and the product and a heat transfer coefficient (α).

The heat that is transferred within a time unit can be simply calculated by Eq. (12.1) [1].

$$\dot{Q} = A \times \alpha \times \Delta T \tag{12.1}$$

The overall heat transfer coefficient α is a function of individual heat transfer coefficients that are the reciprocal of the of present heat transfer resistances [1].

$$\alpha = \frac{1}{\frac{1}{\alpha_1} + \frac{1}{\alpha_2} + \cdots + \frac{1}{\alpha_n}} \tag{12.2}$$

Heat transfer resistances occur everywhere where heat has to be transferred from one media to another (e.g. from the heating medium to a vessel wall (1) or from the wall to the product (3)) but also inside material (e.g. heat conduction through a vessel wall (2) or heat conduction inside the product bulk (4)) (Figure 12.6).

The heat transfer inside the product is mainly dependent on the mixing of the bulk.

At ideal mixing conditions, where the product properties are the same at each location, the heat transfer resistance inside the product is nearly 0 and the heat transfer coefficient is endless.

$$\dot{Q} = A \times \frac{T_H - T_P}{\frac{1}{\alpha_1} + \frac{1}{\alpha_2} + \frac{1}{\alpha_3}} \tag{12.3}$$

That means that the optimum heat transfer for mixing the bulk material is a function of the available heat transfer area, the heat transfer from the heating

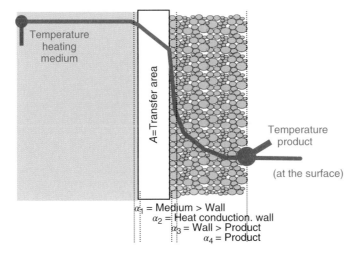

Figure 12.6 Temperature profile during heat transfer to bulk materials.

medium to the wall (1), the heat conduction through the wall (2) and the heat transfer from the wall to the product (3) (Eq. (12.3)).

Typically the overall heat transfer coefficients to friable powders is mainly influenced by the heat resistance from the dryer wall to the product, as the heat resistances to and through the dryer wall are negligible.

12.2.3.2 Product Temperature and Vapour Removal

At real contact drying no inert gas is present inside the process area. The product has to be heated up to the evaporation temperature of the liquid before a phase transition of the liquid into the gaseous phase happens. Assuming ideal mixing, the temperature of the bulk material remains constant until all the moisture on the particle surface is evaporated (Figure 12.7).

The gas inside the process area consists only of vapour from the present solvent. The vapours are removed from the solids because of a pressure difference to the

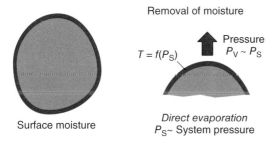

Figure 12.7 Removal of surface moisture.

exhaust point (p_e) of the system.

$$\dot{M} = \frac{1}{c} \times (p_v - p_e) \qquad (12.4)$$

Assuming that the vapour pipe work is dimensioned sufficiently and the flow resistances (c) and also the pressure difference are minimum, the evaporation capacity ($\dot{M}$) is only dependent on the heat transfer to the product ($\dot{Q}$) (Figure 12.8).

Under these conditions the drying process is so called 'heat transfer controlled'.

The bigger part of products that have to be dried in the chemical and pharmaceutical industry have no equal round surfaces. The liquid is captured more or less within capillaries of an amorphous particle structure (Figure 12.9a and b).

For the removal of the vapours that accrue inside the capillaries a pressure gradient from inside the capillaries to the particle surface is necessary. This pressure gradient is developed by heating up the product to a temperature above the

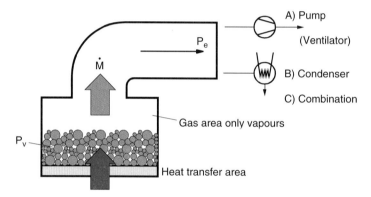

Figure 12.8 Vapour flow.

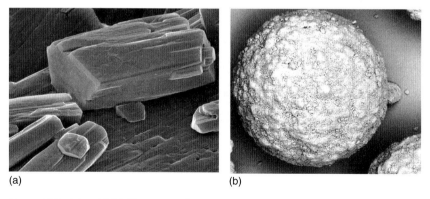

Figure 12.9 (a) and (b) Different particle structures.

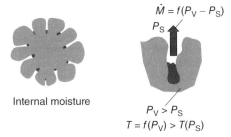

Figure 12.10 Removal of internal moisture.

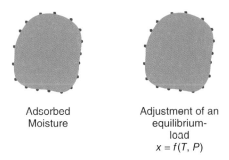

Figure 12.11 Removal of adsorbed moisture.

evaporation temperature. The vapour pressure inside the particles is increased (Figure 12.10).

The drying capacity is no longer only depending on the heat transfer but also on diffusion processes inside the capillaries and so from the structure of the product itself.

Under these conditions the drying process is so called 'residence time controlled'.

A lot of solids are capable to bond the liquid in physically or chemically by adsorption or crystal liquid (Figure 12.11).

The removal of this liquid is only down to an adjustment of an equilibrium load possible that is depending on the final temperature and process pressure. The higher the temperature and the lower the pressure is, the lower the final moisture content.

This form of the drying process is also 'residence time controlled'.

12.2.3.3 Drying under Vacuum

One of the important actuating variables on the drying of products is the evaporation temperature of the present liquid. This temperature is depending on the process pressure (Figure 12.12).

The case that the evaporation temperature is decreasing when the process pressure is lowered gives a lot of advantages especially for the contact drying.

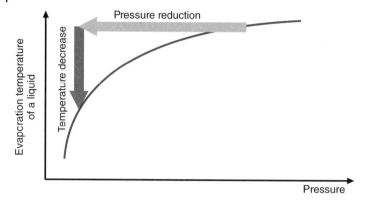

Figure 12.12 Typical evaporation temperature curve.

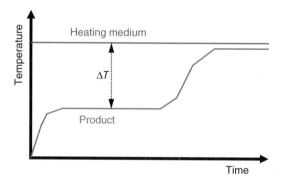

Figure 12.13 Temperature profile during contact drying.

12.2.4
Advantages of the Vacuum Drying

12.2.4.1 Increase of the Drying Capacity

The drying capacity for the removal of the surface moisture is a linear function of the temperature difference (ΔT) (driving force for the heat flow) between the heating medium temperature and the evaporation temperature (Figure 12.13).

At constant heating temperature the evaporation capacity during the removal of the surface moisture is consequently increased under vacuum compared with atmospheric drying because of the higher temperature difference.

12.2.4.2 Gentle Thermal Product Treatment

The quality of many organic products is reduced when the material is processed under too high or too long temperature load. The reduction of the evaporation temperature under vacuum enables a drying with low heating temperatures.

12.2.4.3 Separation of High Boiling Solvents

Contact dryers are typically made out of carbon steel, stainless steel of Ni-alloy materials and are heated using liquid heat transfer fluids via welded on double jackets or half pipe coils.

Typically the maximum operation temperatures of welded constructions and heat transfer fluids are in the range of 320–360 °C.

The evaporation of organic solvents with high boiling temperatures is therefore with standard equipment only possible under vacuum using normal heating temperatures.

12.2.4.4 High Thermal Efficiency

In contact dryers the applied heat is directly used to heat up the product and to evaporate the liquid. Energy losses via hot exhaust gas like at convection drying are not possible. With low evaporation temperatures under vacuum only a small amount of energy is required to heat up the solids. At low heating temperatures the vessel will emit only minimum heat losses to the atmosphere. Hence the required total heat demand is optimized with vacuum drying (Figure 12.14).

12.2.4.5 Processing of Toxic or Explosive Materials

Vacuum dryer installations are closed systems that are ideally operated with no mass transfer from or to the environment.

The gas atmosphere consists of pure solvent vapours. Even in case of slight air leakage there is no risk for an explosion as the vapour density (vacuum) is below critical values.

An emission of toxic products out of the system is also unlikely because of the higher (atmospheric) pressure outside the process area.

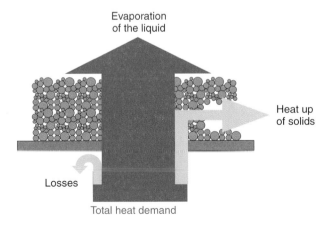

Figure 12.14 Thermal efficiency with vacuum drying.

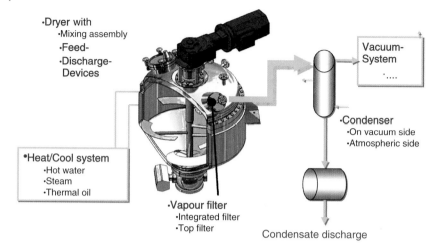

Figure 12.15 Setup of a batch drying system.[2]

12.3
Discontinuous Vacuum Drying

12.3.1
Setup of a Batch Vacuum Drying System

A drying system consists typically out of a dryer, a heating system, a vapour filter, a condensation system and a vacuum system in addition with feed and discharge devices as well as connecting pipework and C & I equipment (Figure 12.15).

The use of vapour filters for the handling of friable powders is essential to avoid a fouling inside the condenser, vacuum pump and pipework by fines. By the use of integrated filter elements any loss of high valuable products is minimized.

12.3.2
Operation of Discontinuous Vacuum Dryers

A typical process sequence for a discontinuous batch vacuum drying process is shown in the following chart (Figure 12.16):

Depending on the product characteristics, the drying is carried out at one constant heating temperature or at different temperature levels.

The heat transfer to wet product is much better than to nearly dry (damp) powders. Hence the evaporation capacity at the beginning of the process is much higher than at the end of the drying. It is often necessary to reach a low pressure level at the end of the drying to make sure that the final moisture can be removed.

---

2) Product Presentation MPE Group GmbH BOLZ-SUMMIX. Simoniusstraße 13, D-88239 Wangen im Allgäu.

12.3 Discontinuous Vacuum Drying

Preparation of the system / inertising	
Filling with product	
Evacuation	
Start heating system	
Drying	Heat up of the product to evaporation temperature
	Removal of surface moisture
	Heat up of product to the final temperature and removal of the internal or absorbed moisture
Cooling	
Break the vacuum	
Product discharge	

Figure 12.16 Typical batch drying sequence.

For the scale up of the peripheral components like vapour filter, condenser and vacuum systems, it is important to use the maximum vapour flows at dedicated pressure levels and not the average values.

In the following chart the influences of different parameters on the drying performance of the equipment are shown.

It is not necessarily the optimum, to design a drying system that is operated only at the lowest pressure and the highest possible temperature, as this would require large and expensive peripheral components (see Figure 12.17). Typically the systems are designed for self adjusting pressure level. That means that the

Parameter	Set point	Influence	Advantage
Temperature	High	High capacity	(+, relative)
	Low	Gentle treatment	(+, relative)
Process pressure	Low	Low evaporation temperature > Low product temperature	(+)
		Low gas density > High volume flows > Large filter areas	(−)
		Low condensation temperature > Use of brines necessary > External condensation >> Large vacuum pumps	(−)
Mixing	High	High heat transfer	(+, relative)
	Low	Gentle product handling	(+, relative)

Figure 12.17 Influences of different drying parameters.

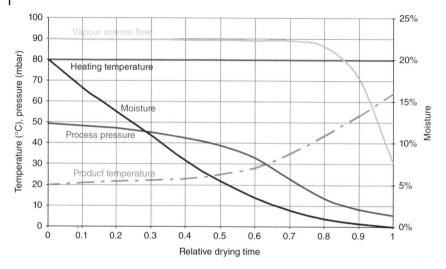

Figure 12.18 Typical drying curve for a batch process.

condenser and vacuum system are suitable to handle high vapour flows at higher pressure levels and the pressure is reduced when the drying capacity is reduced. With such a setup the pressure is low at the end of the drying (see Figure 12.18).

12.4
Continuous Vacuum Drying

12.4.1
Setup of a Batch Vacuum Drying System

The setup of a system for continuous vacuum drying is similar to the setup of batch systems. But in addition dosing devices and feed- and discharge locks are required for a continuous product flow into and out of the system (Figure 12.19).

12.4.2
Operation of Continuous Vacuum Dryers

The major difference between the operation of a batch and continuous dryer is, that the process pressure has to be on one constant level all the time and everywhere inside the process area. With some dryer designs like, for example plate dryers it is possible to realise different temperature zones to optimise the equipment.

As there are always the same conditions and mass flows into and out of the system the design basis for the peripheral components are defined.

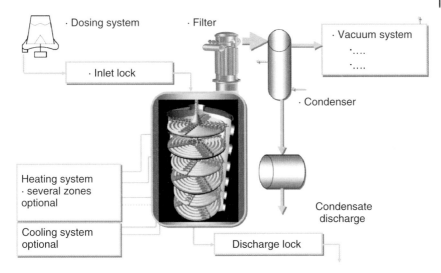

Figure 12.19 Setup of a continuous drying system.[3]

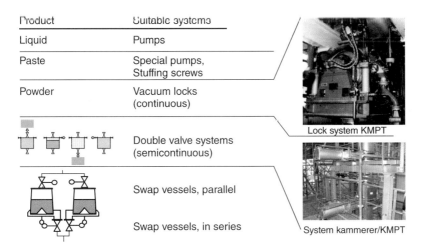

Figure 12.20 Different vacuum lock systems.[3]

12.4.3
Inlet- and Outlet Systems

The selection of suitable feed and discharge equipment for continuously operated vacuum dryers is depending on the product characteristics, the pressure levels and the volume flows. Examples of typical systems are shown in Figure 12.20.

3) Product Presentation KMPT AG. Industriestraße 1-3, D-85256 Vierkirchen.

12.5
Dryer Designs

Vacuum dryers are used for the processing of different type of products under different process conditions and operation modes. Typical applications are the gentle drying of high-value pharmaceuticals in small batches as well as the removal of high amounts of solvents from waste products.

Because of this large variety of possible uses there are a lot different types of dryer designs available on the market.

Figure 12.21 gives an overview about the common dryer types with some basic characteristics.

Type of dryer	Typical application and characteristics	Operation mode
Dryers without product agitation		
Tray dryer 	All product consistencies Manual operation Minimum heat transfer capacity Long drying times	Discontinuous
Belt dryer 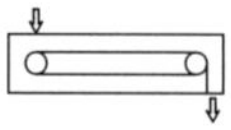	All product consistencies Contact belts Heat input by radiation possible	Continuous
Drum dryer 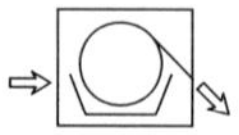	Liquid or pasty products Heated drum or heat input by radiation possible Double drum design available	Continuous
Product mixing by movement of the dryer housing, no internal mixer		
Tumbling dryer 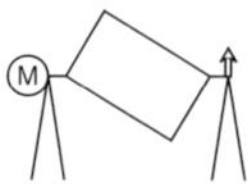	Friable products Gentle drying No mechanical load Poor heat transfer Risk of lump formation	Discontinuous
Double cone dryer 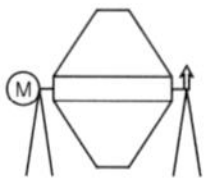	Friable products Gentle drying No mechanical load Poor heat transfer Risk of lump formation	Discontinuous
Product mixing internal mixing assembly		
Conical dryer, screw 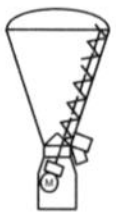	Friable and slightly sticky products Gentle drying Minimum mechanical load Good heat transfer Minimum risk of lump formation Optimum product discharge	Discontinuous

Figure 12.21 Different vacuum dryer designs.

Dryer type	Characteristics	Operation
Conical dryer, Helix type agitator 	Friable products Minimum mechanical load Good heat transfer Good mixing Optimum product discharge	Discontinuous
Pan dryer, fixed or oscillating mixer 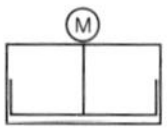	Friable products Simple design Side product discharge	Discontinuous
Paddle dryer 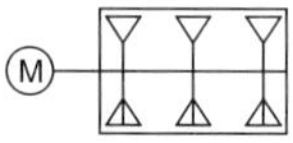	Friable and sticky materials Good mixing but high mechanical load High drive power High amount of cross mixing when operated continuously	Discontinuous, Continuous
Spherical dryers 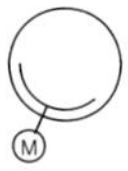	Friable products Good mixing but high mechanical load High drive power Complex design Low filling levels	Discontinuous
Inclined dryers 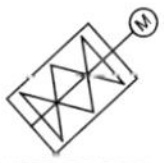	Friable products Good mixing Low filling level Good product discharge	Discontinuous
Mixer kneader dryers, single or twin shaft 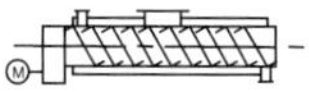	Pasty, crust forming products Good kneading by kneading hooks Self cleaning Long residence times High drive power	Discontinuous, Continuous
Disc dryer 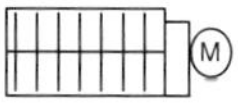	Friable products Large heat transfer areas Risk of cross mixing	Continuous
Thin film dryers, 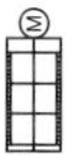	Liquids, pasty or friable products Horizontal or vertical design Good heat transfer Short residence time	Continuous
Plate dryers, 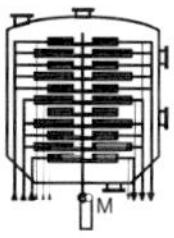	Friable products Large heat transfer areas Different temperature zones possible No cross mixing Homogeneous product quality	Continuous

Figure 12.21 (Continued)

Reference

1. Schlünder, E.U. (1983) *Einführung in die Wärmeübertragung*, 4 Ausgabe, Friedrrich Vieweg & Sohn.

13
Vacuum Technology Bed
Michael Jacob

13.1
Introduction to Fluidized Bed Technology

Using fluidized bed principles, moist bulk materials can be held in suspension and dried at the same time by an air or gas flow (Figure 13.1) [1]. The whole surface of each individual particle is accessible for the fluidization air or gas and very efficient heat and mass transfer processes like drying can be carried out. Accordingly, the treatment of the bulk material is of high quality and the drying duration is short. The high turbulence in a fluidized bed ensures ideal mixing behaviour in the processing chamber, which results in a homogeneous product temperature and a uniform drying of the product [2, 3, 4, 6, 7].

13.1.1
Open or Once-through Fluidized Bed Plants

In open- or once-through fluidized bed plants (Scheme 13.1) [1] fresh air is sucked in, filtered and heated up to the desired process air temperature. While passing through the fluidized bed the heated process air takes up the particle's moisture. Thereafter, the wet process air is exhausted via a filter system. Evaporation rates of up to several hundred kilograms of water or organic solvent per hour can be obtained depending on the size of the apparatus and process conditions (e.g. air or gas flow rate, inlet temperature, product temperature, pressure) [4].

13.1.2
Normal Pressure Fluidized Bed Units with Closed-Loop Systems

Open fluid bed units are not capable of maintaining environmental protection demands when materials to be dried are moistened by organic solvents. Fluidized bed units are therefore equipped with closed duct systems as well as solvent recovery systems (Scheme 13.2). Due to safety regulations (solvent vapour and dust explosion) the entire apparatus needs to be filled with inert gas prior to process

Vacuum Technology in the Chemical Industry, First Edition. Edited by Wolfgang Jorisch.
© 2015 Wiley-VCH Verlag GmbH & Co. KGaA. Published 2015 by Wiley-VCH Verlag GmbH & Co. KGaA.

13 Vacuum Technology Bed

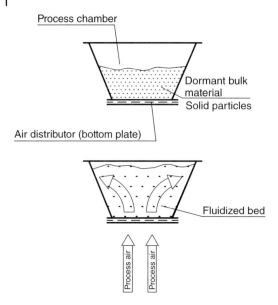

Figure 13.1 Fluidized bed in a conical process chamber.

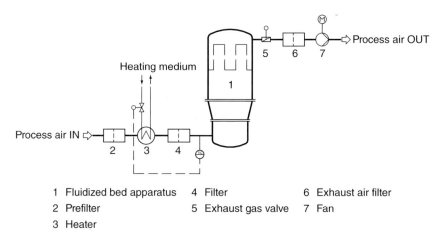

1 Fluidized bed apparatus 4 Filter 6 Exhaust air filter
2 Prefilter 5 Exhaust gas valve 7 Fan
3 Heater

Scheme 13.1 Once-through fluidized bed plant.

begin. The rate of inertisation is monitored by measuring the amount of oxygen during the process. A split stream of the inert gas is channelled to a condenser and cooled down, thusly condensing the solvent vapour and intercepting it into an auxiliary tank. After reheating the inert gas it is again able to evaporate solvent out of the product.

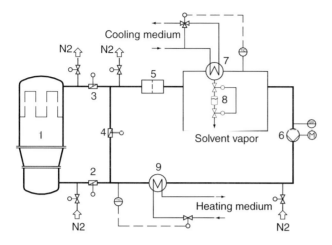

1 Fluidized bed apparatus	4 Modulating valve	6 Condenser
2 Inlet gas valve	5 Filter	7 Intermediate tank
3 Exhaust gas valve	6 Fan	8 Heater

Scheme 13.2 Fluidized bed plant with closed-loop system.

These constructive and control system modifications lead to higher investment costs compared to normal pressure fluidized bed plants. Furthermore the circulating as well as permanent heating and cooling of the inert gas cause rather high operating costs.

13.2
Vacuum Fluidized Bed Technology

The main feature of this unique technique is the operation of a fluidized bed plant at low vacuum conditions. Inertization is not required due to the operation below the minimum ignition pressure. Compared to closed-loop fluidized bed plants, there are many advantages in terms of economic efficiency, ecology and safety aspects:

- considerable reduction of emissions;
- increase of solvent recovery rates;
- lower investment and operating costs.

13.2.1
Layout

In detail, a vacuum fluid bed plant (Scheme 13.3) [1] consists of the following main components:

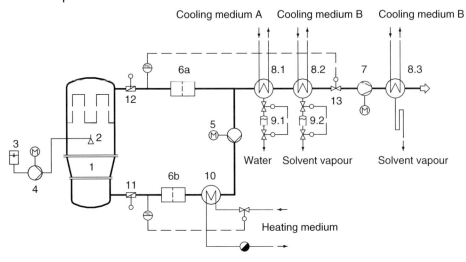

1	Fluidized bed apparatus	8.2	Solvent vapour condenser, level 1, on vacuum side
2	Spraying nozzle	8.3	Solvent vapour condenser, level 2, on pressure side
3	Container for spraying medium	9.1	Intermediate tank for water
4	Proportioning pump	9.2	Intermediate tank for solvent vapour
5	Blower	10	Heater
6a+b	Closed loop filter	11	Inlet gas valve
7	Vacuum pump	12	Exhaust gas valve
8.1	Steam condenser, on vacuum side	13	Vacuum modulating valve

Scheme 13.3 Vacuum fluidized bed plant.

- fluidized bed apparatus featuring inlet air device, process chamber and filter housing;
- closed-loop system with blower, process air heater, filters and duct system;
- solvent recovery system consisting of condensers, vacuum pump and system pressure control valve.

Besides drying processes additional fluidized bed techniques can be performed where liquids are sprayed on the fluidized particles [3, 5, 7]:

- granulation
- spray drying
- coating.

A high-pressure spray pump is necessary for these methods. The spraying pressure typically is between 20 and 80 bar.

Table 13.1 Minimum ignition pressure and ignition temperatures of vapour–air mixtures.

	Minimum ignition pressure (abs) (bar)	Ignition temperature (°C)
Diethyl ether	0.206	170
Hexane	0.32	About 250
Ethyl alcohol	0.44	425
Acetone	0.46	540
n-Pentane	0.52	285

13.2.2
Sequence of Operation

At the beginning of operation a defined operating pressure at low vacuum is set to evacuate the entire machine using a vacuum pump. To ensure a safe operation of the plant without inertisation, the system has to be operated at a pressure below the minimum ignition pressure of the corresponding organic solvent to fulfil explosion protection demands (Table 13.1) [2].

The solvent supplied by the bulk material or atomized into the process chamber using the spray system evaporates due to the temperature and pressure conditions. The process gas and the solvent vapour are carried via the main closed loop through the filter and to the process air heater by the blower. The heated air/solvent vapour mixture is transferred to the process chamber. There it disperses and dries the bulk material. Simultaneously a split stream is diverged from the main gas flow and carried to the recovery system, where most of the condensate precipitates in a vacuum-sided condenser. The remaining condensate is carried off by the vacuum pump to an outside condenser. At the beginning of operation the plant is filled with a blend containing air and solvent vapour, whereas during operation the partial pressure gradually falls because of the vacuum pump. After a short period the amount of solvent vapour inside the machine is nearly 100%. The heat transfer as well as the fluidization itself is done by the pure solvent vapour.

13.2.3
Fluidization at Vacuum Conditions

The development of a fluidized bed depends mainly on the kinetic energy, which is provided by the inlet gas (air or another gas respectively solvent vapour). It also depends on the equivalent differential pressure (Δp) above the fluidized bed [2].

- Δp: gas flow pressure drop caused by the fluidized bed (N m^{-2})
- H_L: bed height at minimum fluidization (m)
- ϵ_L: bed voidage; in this context the ratio of the volume between voids of the particles (where gas flows) and complete bed volume
- ρ_K: density of particle (kg m^{-3})

- ρ_G: density of vorticity gas (kg m^{-3})
- g: acceleration due to gravity.

In case of $Re_K > 100$, the following equation is necessary:

$$u_L = \frac{\delta_K(\rho_K - \rho_\Gamma)\gamma}{X\rho_\Gamma}$$

- C: constant (-)
- u_L: minimum fluidization velocity (m s^{-1})
- Re_K: Reynolds number based on individual particle at minimum fluidization (–)
- d_K: particle diameter (m)

Two factors define the development of a fluidized bed. The first factor is the density of the gas respectively of the solvent vapour. It depends on pressure and temperature. Vapours of common solvents have higher densities compared to air (or nitrogen). As a process gas for fluidized bed operations acetone vapours, for example, have similar properties at 500 mbar (abs) as air at atmospheric conditions. Different solvent vapours such as methanol (standard density 1.43 kg m^{-3}) or toluene (4.11 kg m^{-3}) have little influence on the fluidization. The solvent should not be arbitrary, which inasmuch makes it beneficial.

The second factor is the gas velocity. It is easier to control on one hand; it also has a larger influence on the other hand, because in the equation it is squared. The weight deficits at vacuum conditions can be compensated by higher velocities and with that higher volume flow rates. Duplication can be roughly estimated compared to machines operated at normal pressures. This can be accomplished technologically by a specific blower. The rate of turbulence is equivalent to normal pressure processing.

The fluidization behaviour is mainly influenced by the weight of the bed and the size of the particles. Solid particles with a density of up to about 1.5 kg m^{-3} and a maximum particle size of 3 mm can easily be fluidized.

13.2.4
Heat Energy Transfer under Vacuum Conditions

Unlike the influence of the fluidization behaviour the heat and mass transfer between the process air and the product is highly dependent upon the solvent used. Certain physical properties are the density of the gas at particular pressures and specific heat capacity as well as the essential gas volume. The behaviours of the individual solvent vapours do not differ from one another until below 100 mbar. It is therefore barely possible for acetone vapours to transfer heat energy into the particles at 40 mbar system pressure, no matter which temperatures are obtained in the air heater. Even the radiation of the inlet air duct can hardly be compensated. Then again xylene vapours can be heated under similar circumstances without problems.

This is also valid for spraying purposes, particularly spray drying, in which the considerably high content of solvent has to be removed out of the spraying liquid fast enough to avoid over-moistening.

13.2.5
Applications

The vacuum fluidized bed technique is already being applied for drying, coating and granulating processes at laboratory size as well as for full production scale (Figure 13.2). Relationships and parameters, which have been determined in detailed experimental studies researches stay obtained while up-scaling to production scale [7].

In drying procedures, the system pressure reacts influential on the drying characteristics. Depending on the solvent the effect diverges at high or low system pressures. Critical factors are the heat transferring properties of the fluidization gas or vapour, the solvent's enthalpy of evaporation and the interaction between the solvent and the particle material. To some extent, an immense reduction of the process time can result. For example the duration of a granulation process with subsequent drying of the solvent (acetone) can be reduced from 22 h to 110 min using the one-vessel method with the vacuum fluid bed technique. The method is also used for spray granulation. The properties of the granule and product are primarily influenced by the system pressure. The effects of the inlet air temperature, spray liquid concentration and spraying pressure are similar to the parameters of standard fluidized bed granulating units.

Figure 13.2 (With permission from Glatt Ingenieurtechnik GmbH.)

Furthermore coating processes for applying protective or film coatings (for pharmaceutical purposes, e.g.) to fluidized particles are possible [3, 5, 6].

References

1. GLATT. *Vakuum-Wirbelschicht-Trocknung*, Glatt GmbH, Binzen.
2. Luy, B. (1991) Vakuum-Wirbelschicht – Grundlagen und Anwendungen in der pharmazeutischen Technologie. Dissertation, Universität Basel.
3. Jacob, M., Heinrich, S., and Antonyuk, S. (2013) *Agglomeration von Pulvern in der Vakuumwirbelschicht*, Vortrag, VDI-Fachausschuß Agglomeration, Weimar, March 4–6, 2013.
4. Jacob, M. (2013) *Vacuum Drying in Fluidized Beds*, Vortrag, GVT-University Course Fluidization Technology, Hamburg, November 4–7, 2013.
5. Serno, S., Kleinebudde, P., and Knop, K. (2007) *Granulierung, Verfahren, Formulierungen*. Editio Cantor Verlag, Aulendorf.
6. Luy, B. (1989) Granulieren und Trocknen in der Vakuum-Wirbelschicht; Grundlagen und Erfahrungen bei der Anwendung eines Neuen Wirbelschichtverfahrens. *Pharm. Ind.*, **51** 89–94.
7. Luy, B. and Tondar, M. (1994) Vakuum – Wirbelschicht. *Vakuum Forsch. Prax.*, **6** (2), 123–127.

14
Pharmaceutical Freeze-Drying Systems
Manfred Heldner

14.1
General Information

The drying of products as well as the freezing and subsequent storing at correspondingly low temperatures are well-known processes applied for the conservation of sensitive usually perishable products. These processes are mostly used in the production of foods.

Freeze-drying is a combination of these two processes. With freeze-drying, storage at low temperatures is avoided and in addition, reactions which can still happen at low temperatures if water is present – even in the form of ice – cannot occur. Freeze-drying is the careful drying of a frozen product under vacuum through sublimation of the solvent at temperatures below the freezing point of the solvent.

Freeze-drying is mainly used when water or occasionally other solvents must be removed from a temperature sensitive or structurally difficult product – usually of biological origin – without noticeable influence on the quality of the final product. After removal of the water or solvent, the dry product is then easy to store. It becomes again a product that can be used simply by adding water or the corresponding solvent.

The main areas in which freeze-drying is used can be found in the process technology applied in pharmacy, biology and medicine. A few examples of freeze-dried products are antibiotics, bacteria, sera, vaccines, diagnostic media, biotechnologically produced products and foods.

The main advantages of freeze-drying are

- maintaining the product's original properties (i.e. activity, survival rate of microorganisms, etc.),
- maintaining the structure of the dried product,
- good solubility of the dry product in water,
- possibility of storing the product at ambient temperature.

Vacuum Technology in the Chemical Industry, First Edition. Edited by Wolfgang Jorisch.
© 2015 Wiley-VCH Verlag GmbH & Co. KGaA. Published 2015 by Wiley-VCH Verlag GmbH & Co. KGaA.

14.2
Phases of a Freeze-Drying *Process*

Freeze-drying is a thermal separation process during which the frozen product is dried carefully under vacuum. At the beginning, the product is frozen at atmospheric pressure. After this, in the first phase of drying (primary drying), the water in the product is sublimated. In the second phase (secondary drying), the water still bound in the capillaries is removed through desorption.

The process parameters under which the process is carried out determine the quality of the freeze-dried product.

14.2.1
Freezing

The freezing process influences both the quality of the product and the drying time. The structure of the product created during freezing determines whether freeze-drying is possible and under what technical parameters it can be carried out. Unsuitable freezing will lead to a freeze-dried product with poor solubility, to difficulties in drying and even to the destruction of its biological characteristics as well as poor storage possibilities.

Two parameters are the most important for the freezing process:

- the cooling down rate and
- the required final temperature of the product.

When cooling down an aqueous solution, small ice crystal nuclei begin to form. The water bound in the surrounding area gather on these nuclei. Ice crystals in differing shapes and sizes are formed.

As a rule, when freezing slowly, few crystal nuclei are formed and these grow slowly to form a large structure. Such a structure is favourable for the subsequent sublimation process since it ensures good transportation of the water vapour out of the product to be dried. This in turn leads to shorter drying times. In addition, after sublimation, large crystals leave a relatively loose structure, a pre-requisite for the good solubility in water of the freeze-dried product. When sublimating small ice crystals, with the resulting narrow-pored product structure, the flow of water vapour will be correspondingly slower.

As is well known, the freezing point of water is $0\,°C$. However, the presence of other substances dissolved in water leads to a lowering of the freezing point. For this reason, the required freezing temperature for a product is determined not by the freezing point of water but rather by the eutectic point of the solution to be dried.

Tests to examine the freezing behaviour of the product should be carried out before freeze-drying so that the most favourable freezing method can be determined and then applied during production.

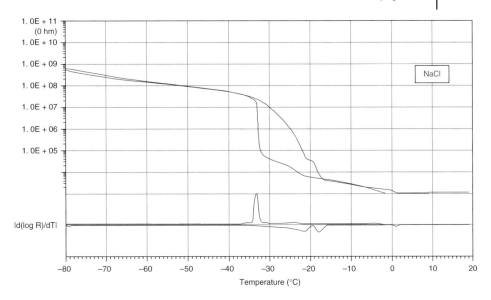

Figure 14.1 Electrical resistance measurement of a sodium chloride solution (NaCl).

Measuring the electrical resistance while at the same time measuring the temperature have proved to be suitable to determine the freezing and melting temperature of the solution to be frozen (Figure 14.1).

As the graph shows, the increase in electrical resistance or the decrease in electrical conductivity during cooling due to the formation of ice crystals. During reheating, the decrease in resistance shows that the ice crystals are melting. These limit temperatures can be determined easily with a low temperature resistance measuring unit.

14.2.2
Primary Drying – Sublimation

The sublimation phase can be described with the help of the phase diagram for water (Figure 14.2). Below the triple point for water (approximately 6.1 mbar and 0 °C), the water to be removed from the product is in the solid or gas state, important for the sublimation process. This is where the technical part of the freeze-drying process begins.

If one follows the entire freeze-drying process in the phase diagram of the solvent (usually water), it is possible to see that the product water is in the liquid state. By lowering the product temperature the water goes into the solid state, ice. If the water vapour is to be sublimated from the frozen product (ice), the total pressure over the ice must be lowered. For pure water, sublimation begins at a pressure of approximately 6.1 mbar which corresponds to the freezing temperature of water of 0 °C. If an aqueous solution is to be sublimated, sublimation can only begin when the pressure is lower than the pressure corresponding to the freezing point of the

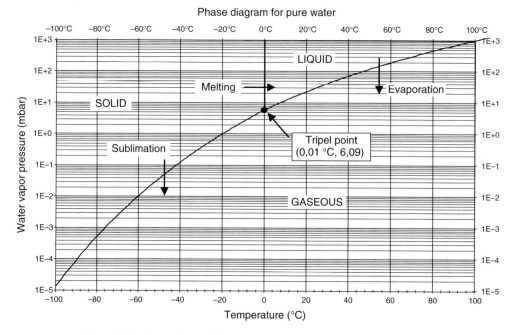

Figure 14.2 Phase diagram for pure water.

solution. The phase diagram thus reflects the connection between the ice temperature and the water vapour pressure. A certain pressure will prevail over any frozen product and this pressure corresponds to the temperature of the product.

During the sublimation process, the temperature in the ice zone of the product must be controlled in such a way that melting cannot occur. At the beginning of primary drying, the water vapour at the surface of the ice in the product is the first to be sublimated. During the further course of drying, the sublimation level recedes into the product so that the water vapour must be transported through the layers of product that are already dry (Figure 14.3). The drying process itself is influenced mainly by the kinetics of the water vapour transportation and by the supply of the necessary sublimation heat.

The supply of heat to the product, necessary for sublimation, is provided by thermal conduction of the gas present in the chamber, through direct contact with the heated product shelves and, to a lesser degree, through radiation. The heat conductivity of gases is dependent on pressure and decreases at low pressures since the number of gas molecules present decreases. It should be noted that the heat conductivity of gases practically disappears at pressures below 10^{-2} mbar. For this reason, in dependence of the melting temperature and the sublimation temperature derived from it, the pressure in the chamber is maintained at the highest possible value during primary drying.

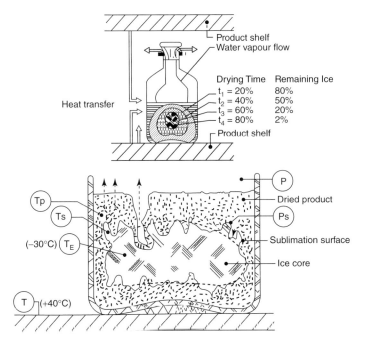

Figure 14.3 Sublimation zones:
T = temperature of shelves and product containers, TE = temperature of the ice core, TP = temperature of the dried product, TS = temperature of the sublimation zones, P = pressure and PS = pressure at the sublimation zones.

Thus, the most important parameters for an economical sublimation process are

- maintaining the highest possible chamber pressure in accordance with the still admissible ice temperature to reduce the drying time;
- very accurate dimensioning of the flow channels for water vapour (i.e. distance between shelves, connections to the chamber and ice condenser, valve cross-sections, etc.) adapted to the required sublimation capacity per time unit and process pressure;
- good heat transfer between the heating surfaces and the product and
- a product layer that is not too thick in the product containers (e.g. vials or product trays).

The data obtained in preliminary tests on the freezing behaviour of the product at certain temperatures, therefore, must be monitored during the entire process and the process itself must be controlled in such a way that the sublimation temperature is maintained throughout. Sublimation can be considered ended when the water frozen as ice crystals has been removed completely from the product to be dried.

14.2.3
Secondary Drying

In many products, the residual moisture in the product after completing primary drying is still too high for optimum storage.

Secondary drying is meant to remove the water adsorbed in the product at the inner surfaces. The freeze-dryer must be designed in such a way that, during secondary drying, the capillary force of the water can be overcome by a greater pressure gradient (e.g. through a very low pressure in the chamber), because in many cases, the product (e.g. proteins) will not accept a considerable increase in temperature.

In addition, the secondary drying process must be controlled in such a way that overdrying of the product is not possible.

14.2.4
Final Treatment

Because of their large inner surfaces, freeze-dried products have the property of absorbing large amounts of moisture from the air very quickly. For this reason, for many freeze-dried products, it is necessary to flood the chamber either with dried air or with protective gas after completing the process. When drying product in vials, these are usually stoppered in the chamber with specially designed lyo-stoppers, which have been placed loosely on the vials before the freeze-drying process is begun (Figure 14.4).

Figure 14.4 Stoppering process of vials within the drying chamber.

14.2.5
Process Control

Control of the process can be carried out very accurately if the freezing and the drying behaviour of a product is known. During freezing, the temperature of the product is measured and the cooling rate carefully controlled until the required final temperature has been reached.

Control of the product temperature using temperature sensors is possible only during freezing and at the beginning of primary drying since it is only during these process steps that sensors are completely immersed in the product. Since the ice around a sensor usually sublimates more quickly than in the other areas, the temperature thus measured cannot be used for a reliable control of the process. If, for example three sensors are placed at three different places in the product, depending on the location of the sensors, different temperature values will be indicated (Figure 14.5). In addition, in a production freeze-dryer – for practical reasons – the product temperature can only be measured in a few vials or product trays. In freeze-dryers with an automatic loading and unloading system, the use of temperature sensors is not possible at all.

A suitable method to measure the temperature during the drying process is to determine the saturation vapour pressure. If the flow of water vapour between the chamber and the ice condenser is interrupted for a short period of time, a saturation vapour pressure in accordance with the water vapour partial pressure diagram will settle in the chamber. The corresponding ice temperature can be taken from the phase diagram (Figure 14.2). This type of measurement is called a

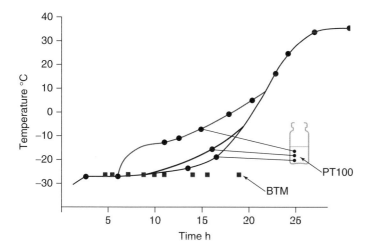

Figure 14.5 Representation of different concepts to measure the product temperature during primary drying: measuring the product temperature with PT 100 sensors and measuring the temperature in the sublimation zones of the product (T_s) by means of 'Barometric Temperature Measurements' (BTM).

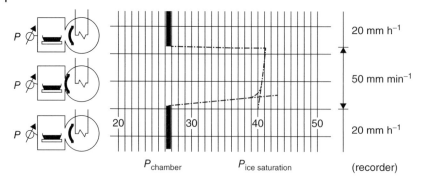

Figure 14.6 Barometric Temperature Measurement (BTM) with evaluation of the process on the recorder strip (schematic representation).

'**B**arometric **T**emperature **M**easurement' (BTM). The time during which the flow of water vapour is interrupted must be so short, that the ice temperature itself cannot increase noticeably.

An important precondition for applying the BTM-method is a very vacuum-tight design of the chamber. This is the only way to ensure that the rise in pressure caused by a possible leak in the chamber remains negligible compared to the rise in pressure caused by the accumulating water vapour. For this reason, the integral leak rate of the drying chamber must be $\leq 10^{-7}$ mbar l s^{-1}.

The end of primary drying can be determined rather accurately with a method similar to the BTM-method. The rise in pressure during a preset time is a direct measure of the amount of ice sublimated per time unit. As the amount of water in the form of ice decreases through drying, the amount of water evaporating also decreases per time unit. If during this step the flow of water vapour between the chamber and the ice condenser is interrupted for a longer period of time (several minutes) and if the pressure in the chamber no longer reaches the saturation value, it can be assumed that the product is sufficiently dry (Figure 14.6).

14.3
Production Freeze-Drying Systems

Freeze-drying systems used in the pharmaceutical industry must fulfil the following conditions:

- The design and manufacture of the system must comply with the requirements set forth by the Federal Drug Administration (FDA) and with current Good Manufacturing Practices (cGMP) guidelines.
- The system must be designed and installed to ensure an optimum production efficiency (e.g. loading and unloading of the product, sterilization, cleaning and clean room conditions).

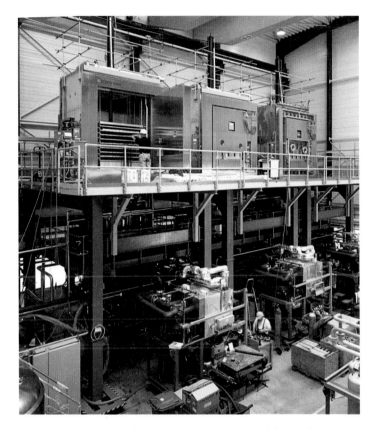

Figure 14.7 Production freeze-drying system designed for installation over two floors.

- Reproducible process and system conditions that can be documented.
- Use of materials that are compatible with the environment; for example for disinfection.
- High product safety using appropriate protective devices, especially when expensive products are to be processed in large product batches The main assembly groups of a production freeze-dryer are shown schematically in Figure 14.7.

14.3.1
Drying Chamber and Shelf Assembly

The drying chamber is made of stainless steel and is usually built into the wall of a sterile room in such a way that only the chamber door is in the sterile room. Inside the chamber is the shelf assembly with the shelves on which the product is frozen and dried. The size of the chamber is determined by the size of the shelf assembly to be installed inside it; the required shelf area is determined by the amount of product to be processed per production cycle.

The following equation is used to calculate the number of vials N per shelf, or of product trays:

$$N = \left(\frac{L}{D} - 1\right) \times \left(\frac{B}{D \times 0.9}\right)$$

L	$=$	Length of shelves or product trays (mm)
B	$=$	Width of shelves or product trays (mm)
D	$=$	Vial diameter (mm).

Example

Shelf dimensions: 1000×1000 mm
Vial diameter: 22 mm

$$N = \left(\frac{1000}{22} - 1\right) \times \left(\frac{1000}{22 \times 0.9}\right) = 44.4 \times 50.5 = 2242$$

This means that approximately 2240 vials per square metre shelf area can be processed.

The design of the drying chamber is influenced by the desired course of production and the installation requirements. If the production process requires a strict separation of the liquid and dry product phases, the chamber must be designed as a pass-through chamber, that is with two doors. In many cases, to reduce the size of the sterile room (sterile rooms are expensive), installation of the system over two floors is requested (Figure 14.7).

The drying chamber is separated from the ice condenser by a valve so that while the ice condenser is being defrosted, the chamber can already be charged again. This chamber-condenser valve is used for BTM and for **Pressure Rise Measurements** (PRM) to determine the end of the drying process.

To freeze and dry products in the chamber, it must be possible to cool and heat the shelves from <-55 to $+80\,°C$. Because of its good thermal properties, silicone oil is used as the cooling and heating medium. It is circulated within the shelf system (Figure 14.8).

The main demands made on the shelf system are

- good temperature distribution over the entire system to attain uniform freezing and drying conditions (maximum admissible temperature difference: $\leq 3\,K$);
- evenness to attain uniform freezing and drying condition;
- high stability for the different product containers (vials) ($P \geq 1.5\,\text{kp cm}^{-2}$)
- High plane parallelism and evenness ($<1\,\text{mm m}^{-1}$) to permit good stoppering of vials; prevention of glass breakage (Figure 14.9).

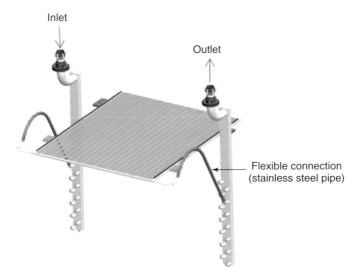

Figure 14.8 Schematic representation of a product shelf.

Figure 14.9 View into a drying chamber with the shelf assembly.

14.3.2
Ice Condenser

The large amounts of water vapour (1 g ice at the usual process pressure of approximately 0.1 mbar has a vapour volume of approximately 10 m^3) which are produced during primary drying (sublimation) can only be pumped off economically by a cold condensation surface.

Dimensioning of the ice condenser is based on the amounts of water vapour to be carried off and the desired operating pressures. The size of the condensation surface must be selected in such a way that the thermal transfer from the ice surface through the growing ice layers to the condenser pipes carrying the refrigerating medium must be as effective as possible. The ice layer should not exceed 2 cm. The condensation surface in the condenser must be designed so that the greatest possible exchange surface is exposed to the water vapour stream.

The geometry of the chamber-condenser valve (Figure 14.10) must be chosen in such a way that there is a good water vapour distribution so that the ice can grow evenly over the condenser pipes installed in the condenser. Temperature control of the condenser pipes is effected over thermal expansion valves; the cooling agent is injected directly into the condenser pipes (evaporator).

The geometry of the mushroom-shaped valve ensures good guidance of the water vapour within the condenser.

When dimensioning the cross-section of the connection between the chamber and condenser and the corresponding valve, it is important to make sure that there can be no restriction of the water vapour flow (Figure 14.10).

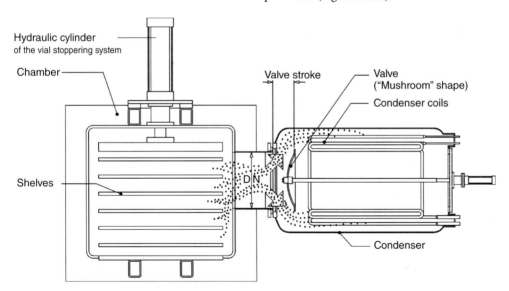

Figure 14.10 Design and installation of the chamber and condenser with a specially designed plate valve.

14.3 Production Freeze-Drying Systems

An example of a calculation of the chamber-condenser valve will illustrate this more clearly:

Total water content in the product (G) (kg)	150
Sublimation time (t) (h)	24
Max. admissible ice temperature in the product (°C)	−30

Note: The maximum admissible process pressure in the drying chamber results from the maximum admissible ice temperature as well as a safety margin of approximately 5 K:

$$P_{process} : 2.5 \cdot 10^{-1} \text{ mbar}$$

A formula developed empirically is used to calculate the maximum sublimation rate (M_{sub}):

$$M_{sub} = \left(\frac{G}{2}\right) \times \left(\frac{1}{\frac{1}{4} \times t}\right) = \left(\frac{2 \times G}{t}\right) [\text{kg h}^{-1}]$$

In the concrete example chosen

$$M_{sub} = \left(\frac{2 \times 150}{24}\right) = 12.5 \text{ kg h}^{-1}$$

The dimensions of the chamber-condenser valve are thus:

- v = specific vapour volume (m³ kg⁻¹) approximately 4500 m³ kg⁻¹ (at process pressure)
- C = The speed of the water vapour transportation (m s⁻¹) lies in the range from approximately 80–90 m s⁻¹ with an efficiently designed system concept
- d = Valve diameter (m).

$$V_{sub} = M_{sub} \times v = 12.5 \text{ kg h}^{-1} \times 4500 \text{ m}^3 \text{ kg}^{-1} = 56250 \text{ m}^3 \text{ h}^{-1}$$

$$D_{valve} = \sqrt{\frac{V_{sub}}{C \times \frac{\pi}{4} \times 3600}} = \sqrt{\frac{56\,250}{80 \times 0.785 \times 3600}} = 0.49 \text{ m}$$

In this example, a valve with a nominal width of DN 500 should be chosen.

14.3.3
Refrigerating System

Mostly two-stage, semi-hermetic refrigerating units are chosen to cool the ice condenser and the shelf heat transfer system. With the presently available low temperature refrigerating agents, R 404 A and R 402 A, the following parameters can be reached:

- ice condenser temperature: $\leq -80\,°C$,
- shelf temperature: $\leq -55\,°C$,
- cooling rate: approximately $1.5\,K\,min^{-1}$

If the product or the process requires lower temperatures, there is always the possibility of attaining these through the use of LN_2.

Advantages for the freeze-drying process of using LN_2:

- High product-cooling rate which can be controlled exactly.
- Shelf temperatures down to approximately $-70\,°C$ possible. A limitation may lie in the increasing viscosity of the heat transfer medium; for example silicone oil.
- Shortening of the secondary drying time through low condenser temperatures (down to approximately $-100\,°C$).
- Low product residual moisture through a low water vapour partial pressure (low condenser temperature).
- High system reliability if there is a power failure (product protection).
- Low investment costs because there is no need of having a mechanical refrigerating system.
- Environment-friendly medium (Figures 14.11 and 14.12).

Figure 14.11 Two-stage compressor refrigeration system.

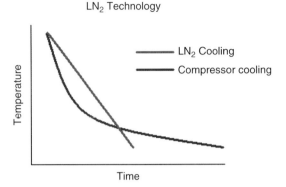

Figure 14.12 Comparison of LN$_2$ cooling performance vs compressor cooling.

14.3.4
Vacuum System

To attain high reaction speeds as well as high water vapour flow speeds in the chamber and condenser, the partial pressure of permanent gases must be lowered to the partial pressure of the water vapour. In order to avoid hindering condensation of the water vapour in the condenser, the share of permanent gases should be low (usually around $\leq 1 \times 10^{-1}$ mbar).

Another criterion when dimensioning a vacuum system lies in the demand that evacuating the system from 1000 down to 10^{-1} mbar should be attained in less than 30 min.

For this reason, in large production systems, three-stage vacuum pumps are used. These consist of two-stage rotary vane pumps and a Roots pump. The greatest advantage of such a pump combination lies in the large volume flow rate at low pressures and in the low ultimate total pressure attained (Figure 14.13).

The attainable ultimate pressure of such a pump set, for example during secondary drying, should not lie lower than the water vapour partial pressure of the condenser when loaded with ice since, otherwise, there is the danger of resublimation of the ice in the condenser or evaporation in the direction of the vacuum pumps. The result would be an increased water content in the pumps and this would lead to inefficient operation.

Example

- A condenser temperature of approximately $-70\,°C$ corresponds to a water vapour pressure of 2.6×10^{-3} mbar
- The required ultimate pressure of the pump set is thus approximately 3×10^{-3} mbar.

Figure 14.13 Vacuum pump system containing three vane pumps and one roots blower.

14.3.5
Cleaning of the Freeze-Drying System

Careful cleaning of the entire system is an important pre-requisite for ensuring the required process and product safety. In the past, cleaning was carried out mostly by hand (wiping, brushing, spraying) while the chamber door was open. The disadvantages of such a procedure are obvious:

- open door in the sterile area;
- very differing individual results and results which could not be duplicated and
- complicated, time-intensive, expensive.

For this reason, concepts were developed for cleaning closed vessels and in spite of their high investment costs, such concepts have become popular.

14.3 Production Freeze-Drying Systems

Such systems are called Clean-In-Place (CIP). They consist of a spray nozzle system installed in the drying chamber and possibly also in the ice condenser and connected to a cleaning system. The spray nozzles are positioned in such a way that all the surfaces in the areas to be cleaned are reached (Figure 14.14).

To prevent contamination, the cleaning water is not in a closed circuit; it is used only once and is thus a lost cleaning agent. The use of demineralised water or water for injection (WFI) at a temperature of 80 °C is usually preferred. To obtain good and effective cleaning, the water should have a pressure of 6–7 bar. In addition, it is important that all ports, lead-throughs and fittings should be designed for easy cleaning. It is also important to make sure that the cleaning water can drain off completely to prevent puddles, dead legs and so on. Since the quality of the surfaces has a strong influence on the cleaning effect, smooth surfaces with a low surface roughness is a precondition for good cleaning.

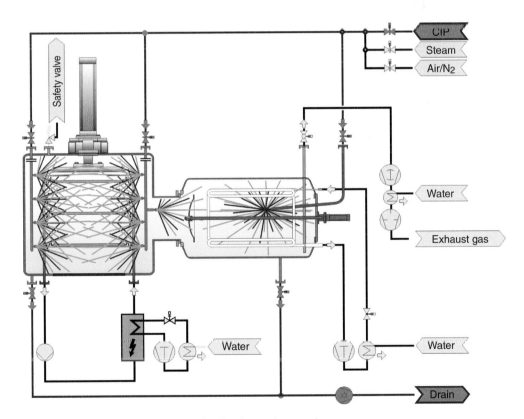

Figure 14.14 CIP cleaning system for the chamber and ice condenser.

14.3.6
Sterilisation

Sterilisation is the elimination or extermination of all micro-organisms as well as the inactivation of viruses that could be found on an object. After sterilisation, such objects must be free of germs capable of reproducing.

At present, there are no specific guidelines or rules for the sterilisation of freeze-drying systems. However, to ensure a sterilisation compliant with cGMP, sterilisation with saturated steam at a temperature of $\geq 121.1\,°C$ corresponding to a saturated vapour pressure of >2.5 bar abs. has become widespread in practice.

Important pre-requisites for good sterilization are

- good steam distribution throughout the system through pre-evacuation of the system (<1 mbar);
- high degree of tightness in the system with very low leak rates to prevent the penetration of non-sterile air during the cooling phase.

Figure 14.15 shows a typical sterilisation cycle. The sterilisation diagram shows that the cycle time is influenced mostly by the heating and recooling of the entire system. In addition, due to the great temperature differences in the system, it is necessary to carry out a leak test after every sterilisation process. The temperature range during a freeze-drying process lies from approximately -80 to $+126\,°C$.

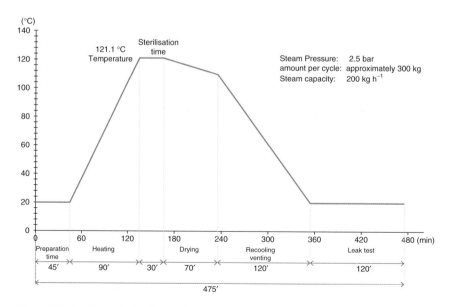

Figure 14.15 Steam sterilisation cycle.

14.3.7
VHP Sterilisation

An alternative to steam sterilisation is the chemical cold sterilisation of surfaces with vaporised hydrogen peroxide (VHP) (Figure 14.16). Hydrogen peroxide vapour is used here as the sterilisation agent.

Hydrogen peroxide vapour is obtained in a generator from a 35% aqueous solution. Such vapour is already very effective at killing spores at the very low concentration of approximately 1500 ppm in the air.

Pre-conditions for effective sterilisation are

- all the surfaces to be dried must be absolutely dry;
- good gas distribution throughout the system;
- uniform temperature of all the parts to be sterilised.

The sterilisation process takes place at ambient temperature.

To attain good distribution of the gas, the freeze-drying system is evacuated to a rough vacuum with its own vacuum pump system. After sterilisation, the H_2O_2 saturated steam is decomposed into water and oxygen with the help of a catalyser. There are no other by-products. For this reason, this process can be considered as completely harmless to the environment.

After sterilisation and before the freeze-dryer can be opened again, it is important to make sure that the H_2O_2 rest gas concentration has fallen to values <1 ppm. For this, several evacuation and venting cycles must be carried out. The entire sterilisation process is validated as required by the OSHA-Regulations. A VHP sterilisation process is shown in Figure 14.17.

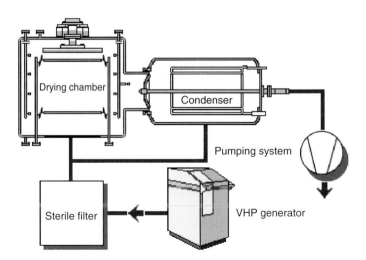

Figure 14.16 Schematic drawing of VHP sterilisation in a freeze-drying system.

14 Pharmaceutical Freeze-Drying Systems

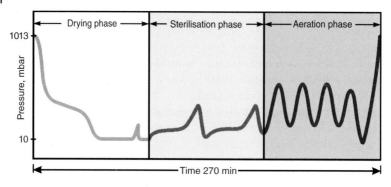

Figure 14.17 Course of a VHP sterilisation cycle.

The main advantages of this type of sterilisation are

- considerably lower investment and operating costs compared to steam sterilisation;
  - No design or acceptance necessary to meet pressure vessel regulations.
  - Very low H_2O_2 consumption per cycle, thus lower costs.
  - No need to have a permanent gas supply.
- sterilisation is effected at ambient temperature;
- short sterilisation cycle of 4–5 h compared to 8–10 h for the steam sterilisation of comparable systems which means a higher plant availability;
- friendly to the environment and not dangerous for operators because the residual products, water and oxygen, are harmless;
- longer life of the freeze-drying system and longer maintenance intervals;
  - No thermal stress to gaskets.
  - No stress through changing pressures as is the case with steam sterilisation.

14.4 Final Comments

This chapter can only show a very general overview of the physical and technical aspects of a freeze-drying process. The decisive factors that determines the quality of a freeze-dried product are always the dimensioning of the different assembly groups and their components, as well as the peripheral processes required for production while keeping economical factors in mind. Thus, the growing demands made in the last few years on sterility and safety against contamination has influenced the product flow also in the periphery. The main problem is to reduce the greatest contamination risk (these are mostly the operators) by using automatic loading and unloading systems. In addition, only such systems can be validated.

An aseptic production precludes applying clean room class 100 conditions. Here, solutions in which filling of the product and the entire product transport are carried out inside isolators (barrier technology) can reduce considerably the enormous costs involved in maintaining clean room class 100 conditions.

This overview shows that freeze-drying – and particularly the peripheral technology which strongly influences the product flow – is subject to continuous improvement.

At present, substituting freeze-drying technology by other processes cannot be envisioned at present, particularly not for the pharmaceutical industry.

Further Reading

Graham, G.S. (1992) Sterilization of isolators and lyophilizers with hydrogen peroxide in vapor phase. International Congress on Advances Technologies for Manufacturing of Aseptic and Therminally Sterilized Pharmaceuticals and Biopharmaceuticals, Basel, Switzerland.

Heldner, M. (1997) Pharmazeutische Gefriertrocknungsanlagen. *Vakuum in Forschung und Praxis*, **4**, 281–288.

Oetjen, G.W., Ehlers, H., Hackenberg, H., Moll, J., and Neumann, K.H. (1961) Temperature Measurments and Control of *Freeze-Drying Processes in Freeze-Drying of Foods*, National Academy of Sciences, Washington, DC, pp. 25–42.

Willemer, H. (1990a) Measurements of temperatures, ice evaporation rates and residual moisture contents in freeze-drying. International Symposium Freeze-Drying and Formulation, FDA.

Willemer, H. (1990b) *Physikalische Grundlagen der Gefriertrocknung*, Klima Kälte Heizung.

15
Short Path and Molecular Distillation
Daniel Bethge

15.1
Introduction

Distillation is a thermal separation technique. It is used to separate liquid mixtures into their various fractions or components by taking advantage of the differences in vapour pressure. The process becomes impractical if the thermal load (temperature, exposure time) becomes too high resulting in chemical reactions and/or decomposition of one or more components. To reduce temperature the distillation is achieved under vacuum. To reduce exposure time at elevated temperature the hold-up in the apparatus is minimized. Nowadays short path and molecular distillation belong to the gentlest distillation technologies.

15.2
Some History

Already 5000 years ago a simple form of steam distillation was used in Mesopotamia to extract essential oils from plants. Greek sailors boiled sea water to obtain potable water. Later mankind tried to increase the strength of alcoholic beverages. Around 1310 Arnold de Villeneuve wrote one of the first known studies about distillation. He describes a pot still as shown in Figure 15.1. The fermented mash is fed into the pot and heated. As alcohol is more volatile than water, alcohol-rich vapours are generated. The vapours are condensed outside of the pot and collected. The distillate has to be redistilled in order to get spirit of a higher degree.

The first successful attempt on continuous distillation and improved efficiency as well as profitability was the Coffey Still in 1831. It was used to produce Whisky. The Coffey Still consists of two columns called analyser and rectifier. The fermented feedstock is preheated against condensing vapours in the rectifier and fed at the top of the analyser. The liquid flows over a copper plate with openings and from there into a down-comer leading onto another, underlying plate, and so on. More than 10 perforated plates (called trays) are typically installed in the analyser.

Vacuum Technology in the Chemical Industry, First Edition. Edited by Wolfgang Jorisch.
© 2015 Wiley-VCH Verlag GmbH & Co. KGaA. Published 2015 by Wiley-VCH Verlag GmbH & Co. KGaA.

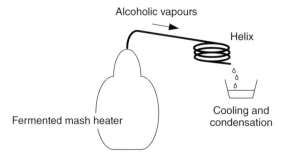

Figure 15.1 Pot still.

Steam is injected at the bottom of the analyser and funnels through the openings upwards. In the created bubbling areas an intensive heat and mass transfer occurs: the less volatile components in the upstreaming vapour condense and the more volatile components of the liquid vaporise. As a result the alcohol content of the downflowing liquid decreases until it is completely stripped off in the bottom effluent. The alcohol content increases in the upstreaming vapours.

The alcohol-rich vapours from the top of the analyser are guided through a vapour duct into the bottom of the rectifier where the alcohol is further concentrated. The top vapours of high proof are liquefied in a condenser and discharged.

15.2.1
Vacuum Distillation

Vacuum science started already in the seventeenth century (Otto von Guericke). With the invention of the diffusion pump (Wolfgang Gaede, 1916) the production of deeper vacuum became technically feasible and economical and, among others, the vacuum distillation technology could develop.

In the production of many chemicals, pharmaceuticals, food ingredients, and so on, as well as in the recovery of valuables from wastes, distillation techniques at ambient pressure are not applicable. Too many undesired reactions and fouling of the apparatus would occur. Examples are oils, fish oil, fatty acids, esters, glycerides, vitamins, waxes, wool fats, mono-/polymers (Figure 15.2).

The boiling temperature can be reduced by applying vacuum. For this purpose a vacuum pump is connected at the vent of the condenser. When reducing the pressure by 1 order of magnitude (e.g. from 100 to 10 mbar) the boiling temperature of most liquid mixtures decreases by more than 25 K.

The applicable vacuum is mainly given by the available cooling medium for condensing of the vapours: an evaporation plant for waste water, for example cannot be operated at below 50 mbar because of the usually available cooling water temperature. Materials with lower vapour pressure can be distilled at lower pressure.

Pot stills and rectification columns can be operated down to a pressure of a few millibars. Lower operating pressures in the evaporator will not be observed even by use of a very powerful vacuum pump. This is due to the pressure drop of

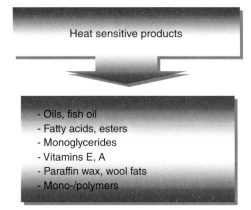

Figure 15.2 Heat sensitive products.

the vapours streaming from the evaporator to the condenser. The pressure in the evaporator cannot be lower than in the condenser plus the head losses.

The head loss of the vapour flow can be eliminated by simply removing the vapour duct. Evaporation and condensation take place in the same chamber. Due to the short path of the vapours this technique is called short path distillation. The principle applies for distillation tasks at ≤ 1 mbar.

Beside the temperature the exposure time is of major importance. Reducing the exposure time by half may have a similar effect as decreasing the temperature by 10 K.

By distributing the feedstock as a thin film on a heated surface the product hold-up in the apparatus and thus the exposure time can be minimised as well. On vertical, heated surfaces the liquid flows downwards due to the gravity whereby the volatile fractions are being evaporated. The produced vapours are condensed on a cold surface installed in a short distance (see Figure 15.3).

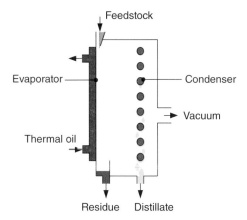

Figure 15.3 The principle of short path distillation.

15 Short Path and Molecular Distillation

In case the vapour density is so low that a vaporised molecule directly hits the condenser, instead of colliding with another vapour molecule, the process is referred to as molecular distillation. Hereto the mean free path length of a vapour molecule must be larger than the distance between evaporator and condenser. Assuming a distance between evaporator and condenser of 50 mm the pressure has to be lowered down to below 0.001 mbar. Only then the mean free path length of a gaseous molecule (diameter 5×10^{-10} m, at 400 K) exceeds 50 mm.

At this low pressure the density of the vapour is small and thus the distillate rate. The relationship can be deduced from the kinetic gas theory and is named according to Langmuir [1]:

$$q = 1577 \times p \times \sqrt{M/T} \tag{15.1}$$

where in q stands for the distillate rate related to the evaporator surface in kilograms per square metre hour, p for the vacuum pressure in mbar, M for the molecular weight of the vapour in grams per mole and T for the absolute temperature in Kelvin.

In practice a fine or high vacuum still is usually operated at ultimate vacuum pressure. The parameters as heating temperature and/or throughput are adjusted in such a way to get the desired distillate rate. The observed vacuum level ensues; it can be derived from the above equation.

Example: At a specific distillate rate of 25 kg/(m² × h), for $\sqrt{M/T} \approx 0.8$ the ultimate operating pressure in the still amounts to 0.02 mbar.

Short path and molecular distillation were developed in the 1930s and 1940s mainly in the UK and in the USA. Different types of apparatus were developed. A still with a disk, rotating at high speed, is named after Hickman, one of the pioneers' [2]. In this machine the centrifugal force is used to distribute the feedstock as a thin film on a heated disk. Exposure time is less than a few seconds!

The feedstock (see Figure 15.4) is fed at a constant rate into the centre of the rotating disk (11). The disk is heated from the rear via heat radiation (9). The volatiles are evaporated and condensed on a water cooled surface (10). The distillate is discharged by means of a pump (6). The residual product flows from the collecting ring surrounding the disk also into a pump (6) for discharge. The cold trap (5) is cooled with liquid nitrogen.

1-Vacuum pump
5-LN2 trap
6-Liquid gear pump
7-Check valve
9-Rotor heater
10-Water cooled condenser
11-Rotor
12-Variable speed pump
P-Vacuum pressure gauge

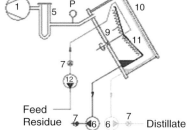

Figure 15.4 Hickman still [3].

15.2.2
Short Path Evaporator

Nowadays the most currently applied apparatus is a wiped film evaporator with integrated condenser (see Figure 15.5). Prior to entering the still the feedstock is preheated and degassed. Feedstock stored at ambient conditions may contain 10 vol% of dissolved gas, that is in 100 l of liquid 10 bar × l of gas! Most of this gas has to be removed prior to a fine or high vacuum distillation. If no vacuum process precedes the short path distillation a degasser applies. The degassed feed is then fed at the top of the cylindrical, heated jacket, equally distributed at the circumference and flows downwards due to gravity as a thin film whereby it is continuously wiped. Due to the heating the volatile components are evaporated and condensed at the internal condenser. The non-volatile fraction, the residue is collected in an inclined ring at the lower end of the cylinder and discharged through the residue nozzle. The distillate is collected at the lower end of the condenser. The non-condensable gas and some vapour are sucked through the condenser and leave the apparatus also at the bottom.

The distillation task determines the required evaporator size. Pilot plant testing is often necessary to quantify. A typical evaporator related feed rate is $100 \, \text{kg}/(\text{m}^2 \times \text{h})$. Depending on the product (viscosity, surface tension etc.) and the goal of the distillation this value might be magnified by up to factor 3 or shrink by factor 0.3. In the latter diffusion limitation (predominant when removing of a little amount of remaining lights) could be the reason.

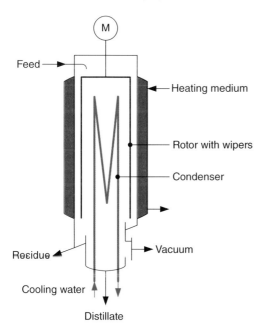

Figure 15.5 Short path evaporator.

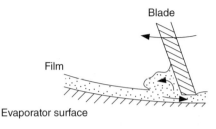

Figure 15.6 Film wiping, top view.

For best heat transfer and circulation the material is distributed and wiped by means of a rotor armed with elements contacting the film. Different types of elements are available.

Blades made from PTFE (polytetrafluoroethylene as appropriate reinforced with glass fibre) or made from steel, pushed by centrifugal force into the film are very often in operation. The film is circulated in a bow wave in front of the blade as shown schematically in Figure 15.6.

Rollers made from PTFE, loosely held on vertical rods and pushed by centrifugal force into the film have excellent self-cleaning properties. The rollers slide over the film surface. Due to the simplicity the system is economic. It is especially recommended for reactive products as polymers.

Blocks made from PTFE and profiled can be used to control the exposure time of typically 3–180 s. Depending on the orientation of the slots the thin film is pushed upwards or downwards.

The rotor can be (totally or partially) equipped with an area covering splash guard (Figure 15.7). In this way any droplet carry-over from the evaporator to the condenser is well avoided.

Special wiping elements as, for example high-speed swinging blades for the drying of products are not commonly used as there were hardly any applications. This may change in the future in the recovery of high boiling valuables from wastes.

15.2.3
The Vacuum System

A cold trap is usually flanged to the vent nozzle of the still. Very light volatiles are condensed here. In many cases the vapours desublimate. Defrosting at regular intervals becomes necessary. The remaining exhaust is sucked off by the vacuum pumps. For an application at 0.02 mbar an oil diffusion pump could be used with a rotary vane pump as backing pump, or three roots blowers in series with an air ejector and a liquid ring pump as backing system. In industrial plants oil booster pumps combined with steam ejectors and liquid ring pumps are widely in operation.

One might expect the bigger the vacuum pump the better. But oversizing may render the system unreliable! Let's consider a medium sized plant to be operated at 0.02 mbar (p_{tot1}). The staff measured a leak rate of 0.05 mbar·l·s^{-1} for the empty

Figure 15.7 Plant assembly/GIG Karasek GmbH; from left: rotor, evaporator (in scaffolding), condenser.

and dry plant. During operation additionally 1 mbar ×l·s^{-1} of (non-condensable) gas has to be removed. The partial pressure of the condensed distillate at the condenser amounts to 0.01 mbar (p_{vap1}). The required effective suction speed at the vacuum nozzle of the still can thus be estimated (sum of gas and vapour load divided by the total pressure resp. gas load divided by the corresponding partial pressure):

$$\frac{(1.05 \text{ mbar} \times 1 \cdot \text{s}^{-1})}{(0.02 \text{ mbar} - 0.01 \text{ mbar})} = 105 \text{ l} \cdot \text{s}^{-1}$$

The stream contains 50% of condensable vapour, that is also 1.05 mbar ×l s^{-1} (vap load$_1$).

At this vacuum level the conductance of the here used cold trap between still and pump amounts to approximately 250 l·s^{-1}. On its way through the cold trap, due to chilling and further condensation, the partial pressure of the condensable vapour decreases from 0.01 to 0.0005 mbar (p_{vap2}). The pressure loss (Δp) is approximately given by the average gas and vapour load divided by the conductance. The vacuum level at the cold trap outlet and the remaining amount of vapour (vap load$_2$) can be estimated by iteration as shown in Table 15.1: index '1' refers to the cold trap inlet, index '2' to the cold trap outlet. The start value of 0.002 mbar for the pressure loss Δp was assumed.

Table 15.1 Iterative estimation of the pressure at the cold trap outlet (p_{tot2}) and of the vapour load.

Iteration step	Gas load	p_{tot1}	p_{vap1}	Vapour load$_1$	Δp	p_{tot2}	p_{vap2}	Vapour load$_2$
	(mbar·l·s)	(mbar)	(mbar)	(mbar·l·s)	(mbar)	(mbar)	(mbar)	(mbar·l·s)
0	1.05	0.02	0.01	1.05	0.002	0.018	0.0005	0.03
1	1.05	0.02	0.01	1.05	0.00636	0.01364	0.0005	0.039954
2	1.05	0.02	0.01	1.05	0.00638	0.01362	0.0005	0.040015
3	1.05	0.02	0.01	1.05	0.00638	0.01362	0.0005	0.040015

The required effective pumping speed amounts to

$$\frac{(1.05 \text{ mbar} \times 1 \cdot \text{s}^{-1})}{(0.01362 \text{ mbar} - 0.0005 \text{ mbar})} = 80 \text{ l} \cdot \text{s}^{-1} = 288 \text{ m}^3\text{h}^{-1}$$

The stream at the outlet of the cold trap contains less than 4% of condensable vapour.

If the installed effective pumping capacity is smaller the desired operating pressure will not be reached. If the actual suction speed is larger the ultimate pressure inside the still will hardly drop further as it is given by the distillate rate. Equilibrium is not reached anymore and the partial pressure of the condensable vapour in the gaseous stream rises. The entrainment of condensable material into the vacuum system increases! Deposits may cause damages.

After having reached the operating pressure it is therefore recommended to limit the vacuum pumping speed using a throttle valve, controlled gas ballast and/or in case of rotating pumps like roots blowers a frequency controller.

15.2.4
Distillation Plant

Besides the vacuum system other peripherals as heating/cooling systems, pumps, control cabinet, and so on are required to operate the still. Figure 15.8 shows a simplified flow sheet of a single stage plant with degasser in front.

Short path distillation is a simple fractionation. The produced vapours do not have contact with the liquid any more. The number of theoretical stages is 1 at a small distillate rate and may rise up to 2.2 at a higher evaporation rate (this is due to the composition change of the thin film along the evaporator height). Although the vapour pressure ratio may increase with decreasing boiling temperature the separation might not be sufficient. To increase the separation efficiency a simple well-known solution may apply:

A first short path evaporator is used to distill off a fraction that meets the quality demands, but at a poor yield; too much of the light fraction remains in the residue. This residue is then distilled in a second still where the light fraction is removed

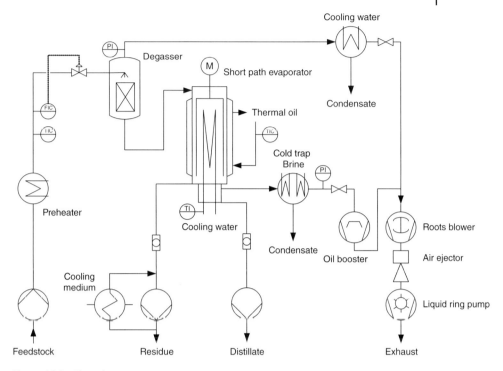

Figure 15.8 Flow sheet.

completely unfortunately together with heavier components. This mixture is then recycled into the feed stream to the first short path evaporator.

15.2.5
Application Examples

Heat sensitive products are distilled under vacuum. Besides the thermal sensibility also the viscosity, respectively the solid content of the material influences the type of deployed evaporator. The different evaporators in their typical application areas are shown in Figure 15.9.

To recover solvents from waste a batch evaporator might do the job (dark grey area in Figure 15.9). Natural or forced circulation evaporators are often used to concentrate products that tend to scale on hot surfaces; the material is heated first and then evaporated by flashing. A falling film evaporator consists of a vertically arranged bundle tube heat exchanger; it allows to process up to huge quantities of material most often waste water as, for example mash from a distillery. The wiped film evaporator (grey) has a very wide field of applications only limited by the investment costs. Equipped with special wiping elements it can also be used as dryer. For gentle treatment the short path and molecular distillation (light grey) apply.

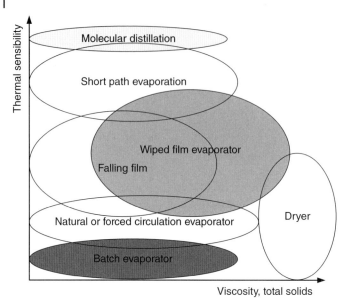

Figure 15.9 Application areas.

In the following are a few application examples for short path distillation:

Lactic acid is a polymer precursor for biodegradable plastic (grocery bags) but is also used in pharmaceuticals, cosmetics and food applications. After the fermentation the water is removed by falling film and/or wiped film evaporation. In the following the lactic acid has to be distilled overhead. This is achieved at about 1 mbar in a short path still. The purity of the lactic acid is of greatest importance. The colour shall remain below a few Hazen even after exposure to heat. To avoid any splashing of liquid material onto the condenser, and thus into the product, the wiper system of the short path evaporator is equipped with special area covering splash guards.

Paraffin wax is required not only for candle-making but there are many other applications as for example coatings of papers or cloths, modification of bitumen, ingredients for cosmetics and even for chewing gum. Wax fractions of distinct properties are produced by short path distillation. The feedstock is distilled in a plant with two (or more) stages: (i) separation of volatiles and (ii) overhead distillation of the product. Apart from the volatiles the feedstock and the product streams are solid at room temperature. Such a plant requires adequate heat tracing of all piping.

An example of paraffin distillation is shown in Figure 15.10. The supposed feedstock consists of 20 components ($C_{10}H_{22}$ ... up to $C_{36}H_{74}$ and some more heavies), each at 5% in the feedstock. The graph shows the composition of a very light fraction (10% of the feed), of the distilled product (distillate) and of the residue.

Monoglycerides are used as emulsifier not only in the food industry (ice cream, bread etc.), in pharmaceutical applications (toothpaste), but also in technical applications. The reaction product is a mixture consisting of mono-, di-

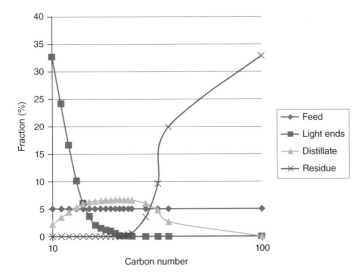

Figure 15.10 Distillation of paraffin (CHEMCAD simulation).

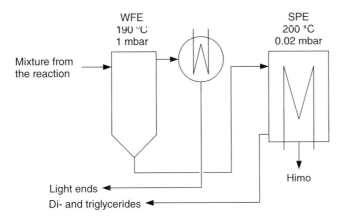

Figure 15.11 Block scheme, concentration of monoglyceride.

and triglycerides. Himo (High concentrated monoglyceride) can be produced by short path distillation. The temperature is still low enough thus avoiding back reactions. The material is degassed first to remove light ends like glycerol and fatty acids. Monoglyceride is distilled overhead in a subsequent step. Di- and especially triglycerides are heavier and remain in the residue. Himo is produced at about 200 °C and approximately 0.02 mbar. The process is shown in Figure 15.11.

Omega-3 fatty acids as DHA (docosahexaenoic acid) and EPA (eicosapentaenoic acid) are known for their wholesome impact on blood circulation. Eskimo people do not often suffer heart attacks despite of the fatty nourishment. This is due to the Omega-3 fatty acids present in the polar fish. As the glycerides cannot be distilled they are first converted into methyl- and/or ethyl-esters. Distillation is achieved at an ultimate pressure of approximately 0.01 mbar and at medium thermal oil heating temperature. Fractions with a high concentration of Omega-3 fatty acids are obtained. After the distillation the esters are converted back into glycerides before being filled, for example into capsules.

Tall oil consists of several components: some moisture and lights, fatty acids, resin acids and heavies. All these components can be further used in different applications. The lights can be separated in a falling film and/or wiped film evaporator. In a consecutive step fatty and resin acids are distilled overhead. The remaining residue solidifies immediately after discharge.

Monomers can be fractionated by short path distillation. To avoid any chemical reactions the heating temperature might be low so that warm water is preferably used.

The purification of **amides** as many other products are performed in a typical two-step operation: removal of light impurities in a wiped film evaporator and overhead distillation of the product in a short path still. Heavy impurities remain in the residue.

15.2.6
New Developments

Research and new developments concern the efficiency, the further reduction of light components in residual products, and the enhancement of throughput.

The removal of a solvent from a valuable product can be achieved by short path distillation. Driving force is the low pressure. By reducing the partial pressure of the already condensed distillate the content of lights in the residue product can further be minimised. This can be achieved by use of chilled water instead of cooling water. Another development consists of moistening or washing the condenser

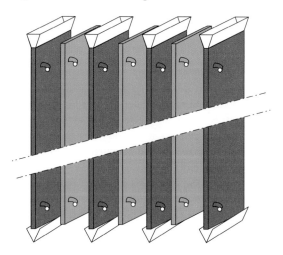

Figure 15.12 Principle of the plate molecular still.

with a liquid which is circulated [4]. The vapours are condensed, mixed in the liquid, and the partial pressure decreases accordingly. The liquid has to be processed before being recycled.

Essential larger quantities of material can be treated with a newly developed plate molecular still [5]. It is a combination of plate-falling film evaporator and short path still. In a vacuum chamber heated and cooled plates are fixed alternately; Figure 15.12 shows the principle. The feed is equally distributed at the top of the heated plates. The produced vapours are condensed on the cold surfaces. At the bottom of the still distillates and residues are collected separately and discharged. Main advantages: no rotating parts and a large evaporation area per volume vacuum chamber.

15.3
Outlook

Short path and molecular distillation always apply in case a gentle treatment of the material is a must, for example in the chemical industry, in the production of pharmaceuticals and of food ingredients or in bio technology. The number of applications in the production of polymers, special chemicals, biofuels and in the recovery of valuables from wastes is increasing.

References

1. Langmuir, I. (1916) The evaporation, condensation and reflection of molecules and the mechanism of adsorption. *Physical Review*, **8**, 149.
2. Hickman, K.C.D. (1947) Commercial molecular distillation. *Industrial and Engineering Chemistry*, **39**, 686.
3. Myers Vacuum, Inc. Pilot-15. http://www.myers-vacuum.com.
4. UIC GmbH Optional Components. http://www.uic-gmbh.de/en/basics/optional-components.html (accessed 8 May 2014).
5. Bethge, D. (2010) Innovative Technik für die Molekulardestillation. *Vakuum in Forschung und Praxis*, **22** (1), 30–31.

16
Rectification under Vacuum

Thorsten Hugen

16.1
Fundamentals of Distillation and Rectification

Distillation is and will remain the premier separation method in the chemical and petrochemical industries. It is the process of physically separating components from liquid mixtures by partial vaporisation and condensation, thereby taking advantage of the differences in component's volatilities. In most distillation columns, vaporisation and condensation are continuously and multiply performed to achieve the desired fractionation effect, and, by convention, this continuous separation method of multiple distillations in one column is referred to as rectification.

Rectification fractionates feed mixtures consisting of two or more components into two or more product streams. These streams include and are often limited to an overhead distillate stream and a bottoms stream, whose compositions differ from that of the feed stream. From the distillate stream, a specific portion called reflux is usually pumped back to the column to increase the product purity. Figure 16.1 provides an overview about the process streams and summarises the main apparatus as well as the peripheral and internal equipment involved in a packed vacuum rectification column.

Rectification usually applies vertical columns, a reboiler generating vapour from liquid and a condenser at the column top for vapour condensation. In case of vacuum rectification, the reduced pressure conditions are usually generated by means of steam jet pumps and/or liquid ring vacuum pumps; for some applications, dry vacuum pumps are used. Inside the column, one or, in rare cases, more liquid phase(s) are flowing down the column and are repetitively and intimately contacted with a vapour phase flowing from the bottom to the top of the column. As long as the phases are not in thermodynamic equilibrium, each phase contact results in mass transfer of components. As a result, the vapour stream is increasingly enriched with the more volatile component(s) before it is removed from top, and the liquid stream is increasingly enriched with the less volatile component(s) on its way down to the column bottom. Following the characteristic of mass transfer, one differentiates the rectification section in the upper column part above the

Vacuum Technology in the Chemical Industry, First Edition. Edited by Wolfgang Jorisch.
© 2015 Wiley-VCH Verlag GmbH & Co. KGaA. Published 2015 by Wiley-VCH Verlag GmbH & Co. KGaA.

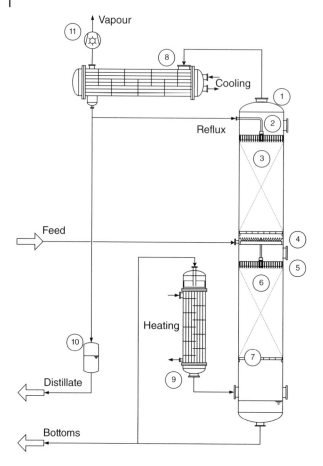

Figure 16.1 Packed vacuum rectification column including internal and peripheral equipment.

feeding location to enrich the lighter component(s) in the vapour phase and the stripping section below the feeding location to concentrate the heavier one(s) in the bottoms [1].

The multiple phase contact inside the column is promoted by internal mass transfer equipment. Three groups of mass transfer equipment are commonly differentiated, which are separation trays, random packings and structured packings. Besides mass transfer equipment, further column internals are required in rectification to ensure the proper operation of the mass transfer equipment. Such internals may include support and hold-down plates, liquid distributors and redistributors, vapour distributor devices, gas–liquid phase separators and liquid collectors that usually do not participate on mass transfer.

Depending on the ease of separating the components and the specific separation requirements related to the desired purity of fractions, more or less separation stages are required to achieve the target fractionation, respectively, in practice,

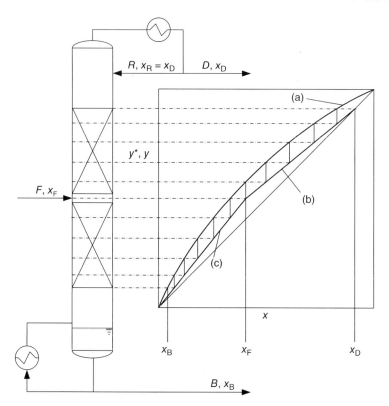

Figure 16.2 McCabe–Thiele graphical equilibrium stage method adapted to packed columns. (a) Equilibrium line, (b) operating line for the lighter component in the rectifying section and (c) operating line for the lighter component in the stripping section. (Adapted from Ref. [2]).

the more or less trays or packing material is installed. The harder the separation of the species, the more separation trays are required in trayed columns, and for packed columns, the more transfer units need to be provided by the packing bed.

The relation between the difficulty in separation and the number of separation stages required can be derived through the McCabe–Thiele methodology. Following this method briefly illustrated in Figure 16.2 [2], the number of separation stages required for the separation task can be derived by stepping off the equilibrium line $y^* = f(x)$ and the operating lines $y = f(x)$ for the rectification and the stripping section of the column in a diagram showing the interrelation between the mole fraction y of the gaseous phase and the mole fraction x of the liquid phase.

The operating lines $y = f(x)$ shown in Figure 16.2 are derived from the respective mass balances of the lighter component in the rectification and stripping sections. The equilibrium line $y^* = f(x)$, however, is derived from the thermodynamic characteristics of the system in terms of the liquid vapour equilibrium. This equilibrium can be then described by the following relation between the concentration x_i of the component i in the liquid phase and its equilibrium concentration y_i^* in the gaseous phase:

$$y_i^* = \gamma_i x_i \frac{\varphi_i^0(p_i^0)p_i^0}{\varphi_i p}$$ (16.1)

φ_i and φ_i^0 are fugacity coefficients describing the behaviour of the species i in the gaseous phase. γ_i represents the activity coefficient of species i in the condensed phase. In ideal systems with similar molecule structures of the components to be separated, γ_i equals 1 and the course of the equilibrium line is similar to the one presented in Figure 16.2. However, in case of non-ideal systems with varying forces between the molecular species to be separated, the activity coefficient becomes $\gamma_i \neq 1$ and the equilibrium line is of a different shape.

16.2
Rectification under Vacuum Conditions

Rectification columns running below atmospheric conditions are referred to as vacuum rectification columns [3]. For industrial applications, pressure levels from several hundred millibar(a) down to 2–10 mbar(a) or even lower are applied. The reasons to perform the process below atmospheric conditions result from the effects of

- lower boiling temperatures and/or
- increased average relative volatilities.

Lowering the boiling temperature reduces thermal degradation effects, fouling, polymerisation and other undesired chemical reactions of bottoms components. In addition, the effects from corrosion are reduced, which significantly proliferate with increasing temperature. Low boiling temperature may also provide more flexibility in terms of reboiler heating medium choice. In comparison to atmospheric conditions, the required reboiler heat exchange area is often reduced under vacuum conditions as a consequence of higher temperature difference between reboiler and heating medium. In some applications using two or more columns, the opportunity to run a column under vacuum conditions makes energy integration of columns possible and energetically feasible.

The second main effect of decreasing the rectification pressure is the increase in average relative volatilities, which is defined as the ratio of the vapour mole fraction over the liquid mole fraction of two components:

$$\alpha = \frac{y/x}{(1-y)/(1-x)}.$$ (16.2)

The average relative volatility can be understood as a measure for the ease of separation. With an increase in average relative volatility, as a rule, the separation gets easier and the number of theoretical stages required to achieve the target fractionation cut is reduced; respectively, at a given number of theoretical stages, the same fractionation split can be achieved at a reduced reflux ratio. The lower the reflux ratio, the smaller the reboiler and condenser duties and so major energy savings can be realised. Providing a given number of theoretical stages and a reflux ratio maintained constant, the product purity is increased with higher average relative volatilities [4].

In practice, the prevention of thermal degradation as well as the increased flexibility in terms of heating medium choice and energy integration are particularly important reasons to run the process under vacuum conditions. However, the advantages gained need to outweigh some unfavourable effects that need to be taken into consideration. One major effect is that the vapour densities are comparably low under vacuum conditions leading to an increase in gas load. Therefore, as the vapour-handling capacities of rectification columns are limited, the inner column diameter usually increases compared to atmospheric conditions, which raises the equipment capital costs. Vapour pipe and valve dimensions also increase leading to additional expenses as well as creating and maintaining vacuum conditions by means of ejectors and/or vacuum pumps. Additionally, compared to atmospheric conditions, the distillate boiling point is reduced and, therefore, low-temperature cooling media are required. Compared to atmospheric conditions, providing a given cooling media temperature, the condenser heat exchange area requirements might be higher as a consequence of a smaller temperature difference between vapour phase and cooling media.

The effects of vacuum conditions on the rectification process are best understood by the study of an example. Exemplary, a rectification process is shown in Figure 16.3 for processing of $10\,000$ kg h^{-1} feed solution consisting of 25 wt% ethanol and 75 wt% water. This binary system shall be separated into an ethanol-rich fraction containing 92 wt% ethanol and a bottom fraction containing less than 0.1 wt% ethanol. Case 1 uses a column operated at atmospheric conditions, whereas Case 2 used two columns, one operated at 2 bar(a) and the other one operated under vacuum conditions at 400 mbar(a). Case 2 represents an energy-integrated system as the vapour from the pressurised column is used to cover the heat requirements of the vacuum column.

As indicated in Figure 16.3, each column of the one- and two-column system is equipped with 30 theoretical stages. The feed location shall be on stage 20 and so the stripping section for removal of ethanol from water is equipped with 10 theoretical stages and the rectifying section for ethanol purification is equipped with 20 theoretical stages. For such configuration, the process key figures presented in Table 16.1 are derived from process simulation and hydraulic design calculations.

Table 16.1 illustrates several characteristic effects from running rectification columns under vacuum conditions. For example, naturally, the bottoms temperature is lowest at vacuum conditions and it increases with pressure, but for this application this effect is of no practical relevance as none of the species is corrosive or subject to thermal degradation at the applied temperature conditions.

In contrast, it is highly relevant that with the same number of theoretical stages, the reflux ratio to achieve the specified fractionation split is lowest for the

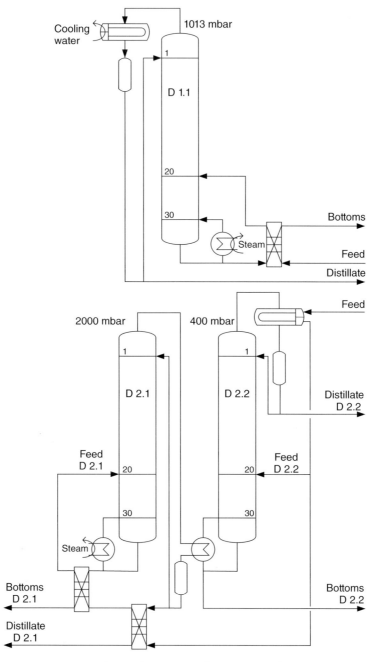

Figure 16.3 One- and two-column systems for ethanol rectification.

Table 16.1 Process comparison of a one- and two-column system for ethanol rectification.

Process Parameter	Unit	D 1.1			D 2.1			D 2.2		
		Feed	Distillate	Bottoms	Feed	Distillate	Bottoms	Feed	Distillate	Bottoms
Ethanol	wt%	25	92	<0.01	25	92	<0.01	25	92	<0.01
Water	wt%	75	8	>99.9	75	8	>99.9	75	8	>99.9
Mass flow	kg h^{-1}	10 000	2600	7400	5000	1300	3700	5000	1300	3700
Pressure level	mbar	1013			2000			400		
Number of theoretical stages	—	30			30			30		
Reflux ratio	—	1.45			1.5			1.1		
Heat consumption	kW	1900			1100			(780)		
Bottom temperature	°C	100			120			76		
Column inner diameter	mm	1000			650			750		
Total pressure drop (packings)	mbar	9			10			9		
Pressure drop per theoretical stage	mbar	0.3			0.3			0.3		

vacuum column. Compared to 1900 kW required for running the process under atmospheric conditions, only 1560 kW (2 × 780 kW) was required for processing of 10 000 kg h^{-1} feed using one column running under vacuum. This effect on heat consumption results from the differences in average relative volatilities of the species under atmospheric and vacuum conditions.

In the heat-integrated two-column system, the total heat consumption is further decreased. As the vacuum column is heated using vapours from the pressurised column, only 1100 kW is consumed in total. Thus, compared to the one-column system, the heat integration of the two-column system leads to a very significant and cost competitive 40% reduction in heat consumption.

Alternatively, for heat integration purpose, one could certainly run column D2.2 at atmospheric conditions to avoid the vacuum system, but in this case, a respective increase in pressure level of column D2.1 were required. The pressure level to be applied for D2.1 must be sufficient to provide an acceptable temperature difference between column D2.1 vapours and D2.1 bottoms. As a result of a higher column pressure, the temperature level of column D2.1 increases and so the availability of a heating medium with a sufficient temperature level is a prerequisite. If this were not available, the opportunity to run column 2.2 at vacuum conditions increases the heat integration flexibility as the bottoms temperature level of column D2.1 can be adjusted by the vacuum pressure level.

The column diameter design presented in Table 16.1 is derived from hydraulic calculations. From these numbers, the effects of vacuum conditions on the column hydraulics can be studied best by comparing the diameter of the pressurised column D2.1 with the one from column D2.2 running under vacuum. Assuming both columns to be packed columns, using Montz wire gauze packings type A3-500 and a gas load limitation equal to F-Factor = 2 (Chapter 16.3) and a pressure drop of maximum 10 mbar over the entire column, one can derive that an inner column diameter of 750 mm was required for the vacuum column instead of 650 mm for the pressurised one as a consequence of the lower vapour density. As a conclusion, the effect of vacuum service on the gas load usually requires a significant increase in column diameter in order not to exceed the gas-handling capacity.

16.3
Vacuum Rectification Design

In the conceptual design stage of a rectification unit, the initial objective is to determine the reflux number and the quantity of theoretical stages required to achieve the desired fractionation split at minimum costs. As a result of this procedure, the heat and mass balances of the column including the internal liquid and vapour flows get known, which are essential for the column design as well as for the selection of the type of internal mass transfer equipment best suited for the specific separation task.

Separation trays and structured and random packings are the prevailing mass transfer internals at choice for rectification columns and, in first deciding between them, a comparative performance design needs to be prepared including examinations on capacity limitations, pressure drop and separation performance. Such examinations require a sound knowledge about the following characteristic parameters and physical quantities to permit process engineering calculations and column design.

16.3.1
Liquid and Gas Load

In terms of the hydraulic design of columns, the liquid and gas loads are major design parameters to be considered. For the liquid flow, the load characteristic is expressed as the liquid volume flow related on the inner column area:

$$B = \frac{V_L}{A_{Col}} [m^3 m^{-2} h^{-1}]. \tag{16.3}$$

The gas load is usually characterised by the F-Factor, which is the vapour capacity factor defined as product of superficial gas velocity and square root of vapour density:

$$F = v_G \cdot \sqrt{\sigma_G} [Pa^{0.5}]. \tag{16.4}$$

16.3.2
Pressure Drop

For every design approach of rectification columns, the pressure drop is an important physical quantity to be considered. The pressure drop is the resistance offered to the vapour flowing through the column and defined as

$$\Delta p = p_{Col,bot} - p_{Col,top} \ [mbar]. \tag{16.5}$$

16.3.3
Separation Efficiency

The number of separation stages required to achieve the desired fractionation is nowadays determined using rigorous methods in commercial software. As a result, the number of theoretical separation stages is determined assuming equilibrium conditions for the heat and mass balances. For packed columns, usually the parameter HETP – height of packing equivalent to one theoretical plate – is used to describe the separation efficiency of structured and random packings and to specify the requirements in packing bed heights to be considered in order to meet the fractionation targets.

With respect to the capacity, pressure drop and separation efficiency, the different mass transfer equipment show specific properties summarised in Table 16.2.

As the gas load capacity (expressed as F-Factor), pressure drop and separation efficiency are not only influenced by the mass transfer equipment type itself but

Table 16.2 Comparison of typical performance characteristics of column packings and trays.

	Type of mass transfer equipment		
	Trays	Random packing	Structured packing
Gas load capacity (F-Factor)	0.3–2.5	0.3–3	0.1 to >4
Pressure drop per theoretical stage	4–10.5	1.2–2.4	0.2–1
Mass transfer efficiency (HETP, mm)	600–1200	450–1500	100–750
Liquid hold up	High	Low to medium	Low

Modified from Ref. [5].

also by, for example the components properties, the column pressure level and the liquid/vapour ratio, the figures provided in Table 16.2 can generally only be understood as indicative figures rather than being universally valid. The figures, however, provide a valuable general guidance to choose the type of mass transfer suited for the respective application and to initiate the column design calculations.

For vacuum rectification applications, and most notably for high vacuum service with low working pressure and a high demand in separation efficiency, there are basically two specific conditions the mass transfer equipment must comply with. The first one is that the gas loads of vacuum columns are comparably high as a result of higher flow velocities of the gas inside the column. Thus, the type of mass transfer equipment used for vacuum applications needs to be capable for high gas loads. The second condition is that the pressure loss over the column height, respectively the pressure drop per theoretical stage, should be as low as possible in order to maintain the advantages gained from reducing the column pressure. For example, for vacuum rectification of a mixture whose relative volatility might be comparatively small already at the column overhead pressure, the relative volatilities become progressively smaller as a result of the rise in pressure by column's internal pressure loss. In other cases, a specific maximum pressure drop must not be exceeded in order to keep the bottoms temperature in an acceptable range in terms of, for example bottom product degradation or corrosion.

Related to the gas loads and the pressure drop, Table 16.2 indicates structured packings to be the best-suited internal mass transfer equipment for vacuum applications. High gas loads, expressed as F-Factor, are tolerated at minimum pressure losses and so it does not surprise structured packings to be the prevailing mass transfer internal used for vacuum rectification applications. In addition, their low liquid hold-ups offer advantages in particular for systems involving thermal degradation with the requirement for very short residence times in zones exposed to heat. And also related to the financial feasibility, structured packings are generally considered cost effective when compared to trays and random packing for such applications [5].

16.4
Structured Packings for Vacuum Rectification

Structured packings are well-established gas/liquid contacting devices used not only in rectification columns but also for stripping columns, absorption and drying towers, air and gas scrubbers and for direct contact coolers. The earliest versions were fabricated from metal gauze followed by sheet, respectively foil metal and expanded metal. Nowadays, structured packings of all sorts of materials are fabricated, such as carbon and stainless steel, hastelloy, copper, monel, tantal and synthetics such as polyethylene, polypropylene, poly(vinyl chloride), polytetrafluoroethylene (PTFE), perfluoralkoxy (PFA) and others. For specific applications, structured packings from ceramics and graphite-filled organic polymers (in case of inner column explosion protection to avoid static electrical load) are also available.

Structured packings are supplied in multifarious shapes. In principle, the packing is composed of geometrically arranged sheets in a regular pattern forming segments. These segments of corrugated elements are assembled to form a packing layer and the packing layers are mounted one above the other to form the packing bed inside the column. In a packing bed shown in Figure 16.4, the elements are stacked such that each subsequent packing layer is rotated 90° with respect to the previous one. Such rotation implies a uniform distribution of both the liquid and vapour phase, which is a major prerequisite of good performance of any structured packing. Maldistribution of liquid is avoided by redistributing the liquid over the entire bed cross-section, whereas the ascending vapour phase is forced to uniformly distribute by sharp changes in flow direction at each transition between the packing layers.

In terms of hydrodynamics, modern column packings are designed for comparably high capacities and use surfaces that promote the separation efficiency. The dominant quantity to which the separation performance parameters are related is the specific surface area defined as packing surface area per packing volume. In general, a smaller specific surface area provides higher capacity with less pressure drop but at the expense of a lower separation performance and higher HETP.

Figure 16.4 Packs of corrugated elements forming a structured packing bed of Montz Type A3-500.

Contrarily, low HETP are achieved from narrow corrugations that usually lead to higher pressure drops per unit height.

The corrugation angle inclination of structured packings is an essential geometry-related parameter influencing both the capacity and the separation efficiency. As illustrated in Table 16.3, well-established and widely used structured packing class uses a standard 45° corrugation angle. In addition, a 60° corrugation angle packing class has also being introduced, which shows a significant increase in capacity. For this modification, however, the percentage capacity gain from increasing the corrugation angle appears to be roughly equal to the percentage decrease in mass transfer efficiency [6].

The target to develop a packing with the efficiency of a 45° and the capacity of a 60° packing was achieved by very effective changes in the geometry of the top and/or bottom ends of the packing elements [6]. As illustrated in Figure 16.5, the Montz high-capacity packing type M, the long and smooth bottom end of each corrugated sheet are bent to end up vertically thereby creating a short transition zone with vertical walls. With this modification, a considerable increase of capacity is achieved with only a very small loss of efficiency compared to the 45° corrugation angle standard [7].

The Montz packing class MN reflects another type of high-performance structured packing that was also developed by geometric optimisation of the corrugation at the lower and upper end of each packing elements. As a result of that optimisation, high numbers of theoretical stages per unit height are achieved at comparably low pressure drops. Table 16.3 summarises the standard and high-performance structured packings offered by established manufacturers exemplary shown for Montz.

For vacuum rectification applications, Table 16.3 indicates the wire mesh packing type A3 to be well suited. This is in particular true for vacuum rectification at a column top pressure level of <20 mbar. The packing class A3 is a woven wire fabric of parallel corrugated elements and is completely self-wetting. Figure 16.6 illustrates the key performance data of Montz A3-500 packing related to pressure drop and separation efficiency.

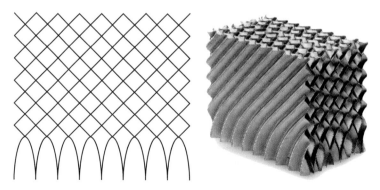

Figure 16.5 Photograph of a Montz high-capacity packing B1-250 M and schematic design illustration [8]. (Permission provided by Julius Montz GmbH.)

Table 16.3 Montz standard structured packings.

	Metal sheet packing	Expanded metal sheet packing	Wire mesh packing
Class designation	B1 (standard) B1.60 (high throughput) M (very high throughput) MN (increased separation efficiency)	BSH (standard) BSH.60 (high throughput) M (very high throughput) MN (increased separation efficiency)	A3 M (very high throughput)
Corrugation angle	45° (standard) 60° (high throughput)	45° (standard) 60° (high throughput)	60° (standard) 45° (on request)
Specific surface (m2/m3)	100, 125, 150, 200, 250, 300, ..., 750		500, 750
Main characteristics	High throughput High flexibility in terms of liquid loading +30% throughput for type M at reduced pressure drop but equal separation efficiency +30% increase in separation efficiency for type MN	Good wettability High fractionation efficiency at high capacity and flexibility +30% throughput for type M at reduced pressure drop but equal separation efficiency +30% increase in separation efficiency for type MN	Low pressure drop per theoretical stage Good wettability through capillary properties, suitable for low liquid loads Applicable for high vacuum service Higher throughput, respectively smaller pressure drop for type M Mainly for vacuum columns with a high demand in separation efficiency and restrictions to acceptable pressure loss
Standard applications	Atmospheric to high pressure columns, partially vacuum columns Absorption columns Direct dryers, refinery columns	Atmospheric to high pressure columns, partially vacuum columns Absorption, drying Dealcoholisation columns (beer, wine)	

16 Rectification under Vacuum

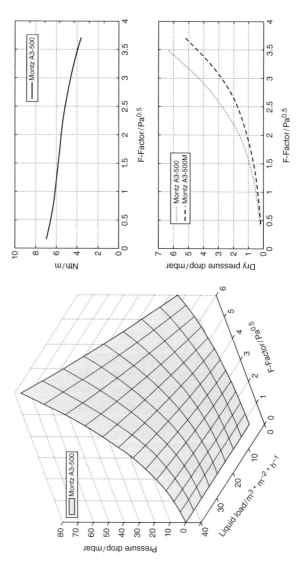

Figure 16.6 Properties of Montz A3-500 woven fibre fabric structured packing.

Basically, the high suitability of woven fibre fabric for vacuum rectification applications results from two characteristic properties:

1) Gauze structured packings with its free surface area of >90% and crimped to 60° from horizontal offer the lowest pressure drop per theoretical separation stage among all packings and thereby, as the pressure at the head of the column is given, it governs the lowest possible pressure at the bottom and thus the lowest temperature, the highest average relative volatility and also the lowest energy consumption.
2) The liquid loads in vacuum rectification is usually much lower than those encountered in rectification under atmospheric pressure or above, and the self-wetting characteristic of woven fibre fabric makes operation at very low liquid volume flows possible.

Besides these capacity- and pressure-drop-related advantages, further beneficial characteristics include the high flexibility in terms of liquid and vapour loads and a high efficiency being essentially independent from column diameter.

For efficient operation of vacuum rectification columns using woven fibre structured packings, some technical performance features related to the interaction of gas, for example liquid load, pressure drop and separation efficiency, are to be considered with care. As indicated in Figure 16.6, a rapid increase in pressure drop with only a slight increase in gas rate develops at flooding conditions respectively close to them. The higher the liquid load the lower the gas load at which this characteristic behaviour applies. Related to mass transfer, each increase in gas rate at that point is followed by a rapid decrease in separation efficiency and, therefore, as a rule for efficient operation, the optimum load factor should be equal to a vapour capacity factor that in the vacuum systems is not higher than 70–80% of the corresponding flood point. At such loads, the woven fibre fabric structured packing shows performance characteristics superior to alternative available mass transfer devices for vacuum rectification.

The ultimate performance of all types of structured packing including the woven fibre ones, however, not only depends on the loading characteristic, the specific surface area and its geometrical properties but also highly affected by, for example the column diameter, the system properties, component concentrations, pressure level and the depth or height of the packing bed. Thus, for a safe design of packed columns it is recommended to apply a 'factor of safety' related to the specific separation efficiency of the packing, to be applied when component separation is important.

Nomenclature, Applied Units

A	Area, m^2
B	Liquid load $m^3\ m^{-2}\ h^{-1}$
F	F-Factor, $Pa^{0.5}$
HETP	Height equivalent to a theoretical plate, m
p	Pressure respectively partial pressure, Pa

PFA	Perfluoralkoxy
PTFE	Polytetrafluoroethylene
Nth	Number of theoretical plates
V	Volume flow, m^3 h^{-1}
x	Mole fraction of liquid phase
y	Mole fraction of gas phase

Greek Symbols

α	Average relative volatility, $\alpha_{ij} = p_i^0/p_j^0$
φ	Fugacity coefficient
σ	Mass density, kg m^{-3}
ν	Superficial velocity, m s^{-1}
γ	Activity coefficient

Subscripts and Superscripts

B	Bottom product (bottoms)
bot	Bottom
Col	Column
F	Feed
D	Overhead product (distillate)
L	Liquid
R	Reflux
i	Component i
j	Component j
top	Top
*	Equilibrium state
0	Pure

References

1. Seader, J.D. and Henley, E.J. (2006) *Separation Process Principles*, John Wiley & Sons, Inc.
2. Grote, K.-H. and Feldhusen, J. (2004) *Dubbel – Taschenbuch für den Maschinenbau*, Springer, Berlin, Heidelberg, New York.
3. Thomas, C.E. (2011) *Process Technology Equipment and Systems*, Delmar, New York.
4. Kister, H.Z. (1992) *Distillation Design*, McGraw-Hill, Inc., New York.
5. Ludwig, E.E. (1994) *Applied Process Design for Chemical and Petrochemical Plants*, Gulf Publishing Company, Houston, TX.
6. Olujić, Ž., Seibert, A.F., and Fair, J.R. (2000) Influence of corrugation geometry on the performance of structured packings: an experimental study. *Chemical Engineering and Processing*, **3**, 335–342.
7. Olujić, Ž., Seibert, A.F., Kaibel, B., Jansen, H., Rietfort, T., and Zich, E. (2003) Performance characteristics of a new high capacity structured packing. *Chemical Engineering and Processing*, **42**, 55–60.
8. Olujić, Ž., Jansen, H., Kaibel, B., Rietfort, T., and Zich, E. (2001) Stretching the capacity of structured packings. *Industrial and Engineering Chemistry Research*, **40** (26), 6172–6180.

17
Vacuum Conveying of Powders and Bulk Materials

Thomas Ramme

17.1
Introduction

In the chemical industry vacuum conveying is mainly used for the transport of raw materials and additives which are required for the various processes and treatments. Transferred are bulk materials like powders, dusts and granules and sometimes even viscous products or liquids.

Similar conveying tasks are to be found in the production steps of making pharmaceuticals and around food processing. In all these cases the material to be conveyed is picked-up at a certain location in the plant by vacuum. After transportation over a given height and distance the powders and bulk materials are discharged and fed actively at the desired destination.

Based on the various conveying tasks, there is a large number of possible suction (pick-up) and discharging locations. (Figures 17.1 and 17.2a and b).

A look at the above-mentioned configuration possibilities lead to a first determination of the parameters given by the layout of the plant, especially the conveying height and total conveying distance.

In addition, special attention should be paid on the characteristics of the specific materials which are to be conveyed (Figure 17.3). Moreover the parameters are interconnected if, for example the influence of the bulk density on the conveying line diameter is considered.

Figure 17.4 shows the main parameters which should be investigated if a vacuum conveying system is configured. First the general conveyability must be checked where in a second step the throughput ($kg\,h^{-1}$) required by the process has to be confirmed.

Due to the large number of parameters and often difficult to measure properties, given by the huge range of powders and bulk materials, the standard method for selecting the right type of vacuum conveyor is to conduct 1 : 1 scale vacuum conveying trial.

Vacuum Technology in the Chemical Industry, First Edition. Edited by Wolfgang Jorisch.
© 2015 Wiley-VCH Verlag GmbH & Co. KGaA. Published 2015 by Wiley-VCH Verlag GmbH & Co. KGaA.

Material flow in a vacuum conveying system with feeding and discharging locations
Various pick-up and loading options

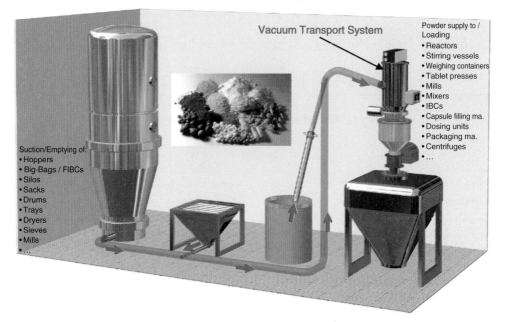

Figure 17.1 Application examples for vacuum conveying. (Volkmann Vakuum Technik GmbH.)

17.2
Basic Theory

17.2.1
General

Basically the same physical principles apply for vacuum conveying as well as for positive pressure conveying. It involves multiple phase flow [1] where the bulk material (solid phase) is moved by the air (gas phase).

With positive pressure conveying the material to be conveyed is fed into the conveying line at one central point and can be transported to one or more destinations. In return vacuum conveying allows the pick-up of materials by suction at various locations and the product is fed into one central destination point. For example, different raw materials are picked up from Big-Bags by vacuum and are transferred into one central mixing- and weighing-container.

At positive pressure conveying often high pressures are applied, whereas for vacuum transfer only approximately 1013 mbar is available and usually only a pressure difference of maximum 900 mbar is applied. This is the reason why

Reactor loading in API plants & chemical industries

INEX = Nitrogen inert system

(a)

(b)
Pressure rated vacuum conveyors / powder locks

Figure 17.2 (a,b) Vacuum conveying for powders and bulk materials in chemical processes. (Volkmann Vakuum Technik GmbH.).

Chemical	Pharmaceutical	Food	Divers
TiO_2	Laxative granule	Icing sugar	Toner powder
Flame soot	Paracetamol powder	Cocoa	Aluminium dust
Sulphur	Pharma capsules	Garlic powder	Blood powder
SiO_2	St. John's wort (drug)	Apple sauce	Al_2O_3-powder
Talcum	Acetylsalicylic acid	Cream-fat-powder	Magnesium swarfs
Zinc oxide	Ascorbic acid	Bacon pieces	Casting powder
Silicon carbide	Sterile caps	"Milk"	Mussel shell
Bentonite	Cellulose powder	Wheat starch	PE-granule
Activated carbon	Plant drugs	Desiccated coconut	Strontium ferrite powder
Resin granule	Barium sulphate	Mixed spices	Colour pigments
Sodium cyanate	Magnesium powder	Chanterelles	Palladium ashes
Ethylene-poly.	Tablets	Gelee granule	Electro waste (shredded)

Figure 17.3 Examples for conveyed materials in various branches of industry.

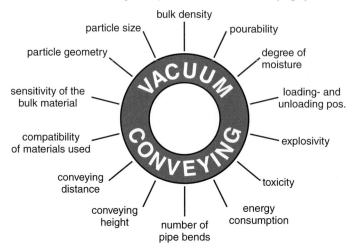

Figure 17.4 Process and material parameters.

in the chemical industry positive pressure conveyors are mainly used for long distance conveying and high throughput rates. Vacuum conveying involves an easier feeding of the material (by aspiration) which takes place at atmospheric conditions.

At positive pressure conveying, the powder has to be fed against a positive pressure with special valves (powder lock) which in general increases the costs. The usually hot blowing air at the feeding point can cause further problems or requires further treatment. Leakages of the downstream conveying line, cyclones and filters can cause environmental dangers, especially if very fine powders like toner or soot are conveyed under high positive pressure.

Because of the vacuum, it is easier to design a dust free, closed system (containment) with vacuum conveyors. Rising dust is sucked into the vacuum line. For some positive pressure conveying systems, it can be necessary to have a pressure-rated sending vessel at the feeding point and in addition a cyclone and filter arrangement above the material's destination. With vacuum conveying only one receiving vessel with integrated cyclone and filter is needed. No matter which type of pneumatic conveying is applied, for a theoretical approach the conditions in an empty pipe and pure air are considered. The pressure loss Δp_L [2] is than given by Eq. (17.1).

$$\Delta p_L = \lambda_L * \frac{\Delta l}{d} * \frac{\rho_L}{2} * w^2 \tag{17.1}$$

with

λ_L	=	pipe friction factor (air flow)
Δl	=	pipe length in metre
d	=	pipe diameter in millimetre
ρ_L	=	air density in kilogram per cubic decimetre
w	=	air velocity in metre per second

The dimensionless pipe friction factor λ_L is determined in dependence of the Reynolds number Re (Eq. (17.2)) and the relative pipe roughness k/d. The literature [3] shows diagrams.

$$\mathrm{Re} = \frac{w * d}{\upsilon} \tag{17.2}$$

υ = kinematic viscosity of air in square metre per second.

The equally dimensionless relative pipe roughness shows the absolute roughness k (in millimetre) with regard to the internal pipe diameter d.

Apart from the pressure loss of the empty pipe and further insertions like valves or pipe bends, an additional pressure loss is created by feeding product in the conveying line. The vacuum generator must be able to overcome this total pressure loss and at the same time must be able to generate the necessary velocity (airflow rate with regard to pipe diameter) for the movement of the material to be conveyed. Basically the necessary air velocity for conveying can be found by considering the suspense velocity of one single particle. From the force equilibrium acting on a round particle in a vertical air flow the suspense velocity v_S is given in Eq. (17.3).

Gravity velocity = lifting force + resistance force

$$v_S = \sqrt{\frac{4}{3} * d_K * \frac{g}{c_w} * \left(\frac{\rho_K}{\rho_L} - 1\right)} \tag{17.3}$$

with

d_K	=	corn diameter in millimetre
g	=	gravity acceleration metre per second squared
c_w	=	drag coefficient
ρ_K	=	corn density in kilogram per cubic decimetre

17.2.2
Typical Conditions in a Vacuum Conveying Line

As seen in Section 17.2.1. the total pressure difference during powder conveying depends on the amount of material in the conveying line. Basically at vacuum conveying three different types of conditions can be described (Figure 17.5): dilute phase, dense phase and plug flow conveying [4].

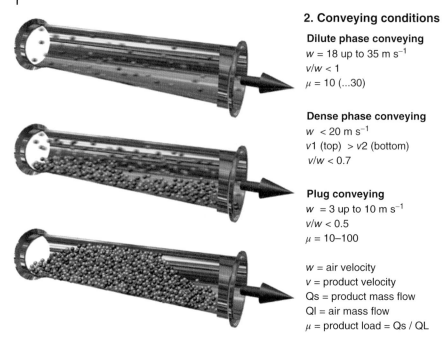

2. Conveying conditions

Dilute phase conveying
w = 18 up to 35 m s^{-1}
v/w < 1
μ = 10 (...30)

Dense phase conveying
w < 20 m s^{-1}
v1 (top) > v2 (bottom)
v/w < 0.7

Plug conveying
w = 3 up to 10 m s^{-1}
v/w < 0.5
μ = 10–100

w = air velocity
v = product velocity
Qs = product mass flow
Ql = air mass flow
μ = product load = Qs / QL

Figure 17.5 Overview of the different conditions in a conveying line.

17.2.2.1 Dilute Phase Conveying

At dilute phase conveying the air velocity w is much larger than the suspense velocity v_S and can be measured from approximately 18–35 m s^{-1}.

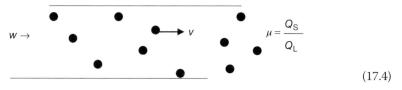

$$\mu = \frac{Q_S}{Q_L} \quad (17.4)$$

The ratio of product velocity v to air velocity w is smaller than 1. The product load μ (Eq. (17.4)) which represents the ratio between the product mass flow Q_S and the air mass flow Q_L is around 10 but in dependency of the material to be conveyed can reach 30.

Depending on the characteristic of the vacuum generator (air-flow rate above vacuum level) dilute phase conveying leads to high throughput rates. However the high velocity may lead to wear or damage of the material to be conveyed.

17.2.2.2 Dense Phase Conveying

If in a horizontal or declined conveying line the air velocity w is reduced under 20 m s^{-1} the material starts to drop down in the lower half of the pipe and moves like a string. The formation of this string is strongly depending on the material. Plugs may occur and settle for a short time in the lower half of the pipe where above that dilute phase conveying takes place due to the velocity increase by the diameter reduction.

The ratio of product mass flow v to the air velocity w is smaller than 0.7. The product load μ is larger than with dilute phase conveying. Dense phase conveying is a gentle type of conveying and requires less energy.

17.2.2.3 Plug Flow Conveying

With further increased product load μ and reduced air velocity w single plugs occur in the conveying line. Downstream the conveying line they dissipate and build up again constantly. Therefore the air velocity w can even in vertical conveying line be reduced to values below the suspense velocity v_S of a single corn. This leads to a gentle conveying. The vacuum generator must be able to create a high vacuum. Simple blowers cannot be used for this type of conveying. Plug flow conveying is not only applied for powders and granules, but also for viscous and liquid products.

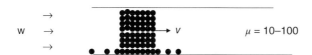

The air velocity w is between 3 and 10 m s^{-1}, whereas the ratio of product velocity v to air velocity w is smaller than 0.5. The product mass flow can be a hundred times larger than the air mass flow.

The required power for the vacuum generator can roughly (incompressible) calculated with Eq. (17.5).

$$P = \Delta p * \dot{V} \qquad (17.5)$$

with

P	=	power in kiloWatt
Δp	=	pressure difference in millibar
$\dot{V}$	=	air-flow rate in cubic metre per hour.

The energy requirements for dense phase conveying are comparable with dilute phase because although the required air flow rate is much lower a larger pressure difference is required. The ratio is proportional.

17.3
Principle Function and Design of a Vacuum Conveying System

In the last decades, different designs of vacuum conveying systems were invented. They have in common the basic components (Figure 17.6), consisting out of three main groups:

1) vacuum generator
2) conveying- and receiver vessel
3) filter system.

The *vacuum generator* supplies the necessary negative pressure and air-flow rate which allows the suction of the bulk materials through the conveying line in the *separator and receiving vessel*. In the lower section the conveyed material is collected. The *filter* prevents that small powder particles appear in the exhaust air of the system. If the filling volume is reached, the vacuum is switched off rapidly and the *discharging valve* opens. Simultaneously the filter is cleaned. If the discharging cycle is finished, the discharging valve is closed again and the vacuum starts rapidly. A new suction cycle takes place.

For the *generation of the vacuum* distinctive principles are applied with regard to the properties of the countless different bulk materials and the desired conditions in the conveying line:

- side channel blowers (in suction mode),
- oil-sealed sliding vane pumps,

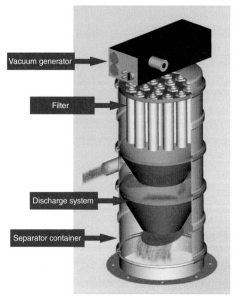

Principle design of a modular

dense phase/plug flow

Vacuum Conveyor

Conveying cycles:

1. Suction/vacuum
2. Filling
3. Aspiration/filter cleaning
4. Discharging

Figure 17.6 Functional model of a vacuum conveyor. (Volkmann Vakuum Technik GmbH.)

- multiple-stage, compressed air-driven vacuum pumps,
- liquid ring vacuum pumps,
- roots blowers (in suction mode)/rotary lobe pumps,
- ejectors.

Side channel blowers generate only a small negative pressure and therefore are only suitable for the conveying of easy-flowing granules and short conveying distances. Oil-sealed sliding vane pumps can be used for difficult powders and for longer distances.

However with the multiple-stage, compressed air-driven vacuum jet pumps, all types of conveying are possible, because high air-flow rates for dilute phase conveying as well as large pressure differences for plug flow conveying can be generated.

In a similar variety *conveying and receiver vessels* are applied. Such are

- single containers,
- single containers in pressure-rated design,
- modular containers,
- expanded modular systems with interchangeable elements and
- separators out of special materials: for example, stainless steel of various grades and alloys, glass coatings or special treatments of the internal surface.

This selectional concept applies in the same way to the *filter systems*. Following typical configurations for applications around the chemical industry are examined more in detail.

17.3.1
Multiple-Stage, Compressed-Air Driven Vacuum Generators

The vacuum generator must supply sufficient air-flow rate and must overcome the resistances (pressure losses) given from the system by creating the appropriate vacuum level. Coming from the various principles of creating a vacuum, especially the multiple-stage jet pumps have specific advantages for small and medium size vacuum conveying systems (throughput rates up to $10\,t\,h^{-1}$, often much less). These capacities are regularly required for typical conveying tasks in chemical processes (Figure 17.7).

The driving gas p+, which normally is compressed air, enters the pump and flows through the first nozzle stage. This first nozzle stage consists out of special laval nozzles (Figure 17.8).

The entry part of the laval nozzle is shaped like a norm nozzle. In this subsonic area of the laval nozzle the compressed air is strongly accelerated which leads finally to sonic speed (w_S) at the tiniest diameter.

The changing of the characteristic of the flow when reaching sonic speed is used in the now-following diffuser. Here, the increasing diameter leads to a further gain in velocity at supersonic speed level. The supersonic air velocity reaches a peak at the outlet of the nozzle and in correlation a strong pressure drop takes place

Principle design of a multiple-stage, gas-jet driven vacuum pump

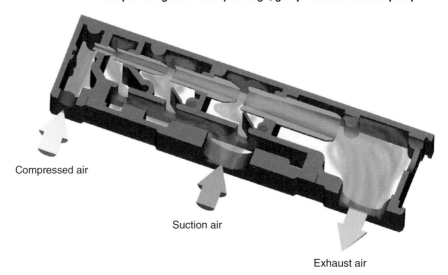

Figure 17.7 Functional model of a multiple-stage jet pump.

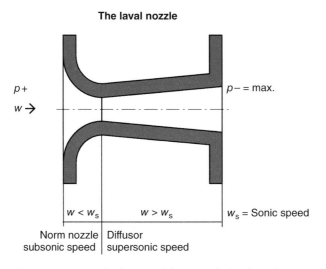

Figure 17.8 Principle design and function of a laval nozzle.

which produces an extraordinary high vacuum. This vacuum level is important, if later plugs are to be conveyed through a conveying line. The now-following three- or four-nozzle stages recycle the still-present high air-flow energy in order to increase the total suction capacity of the pump. These following stages feed themselves and are not requiring any further external energy supply. The high air-flow

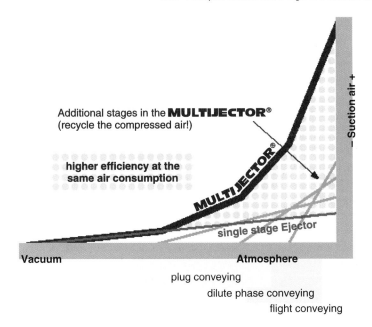

Figure 17.9 Typical characteristic of a multiple-stage ejector.

rates are necessary to accelerate the material in the conveying line and allow dilute phase conveying if necessary (Figure 17.9). Especially around powder and bulk material conveying this principle has specific advantages. First of all, the multiple-stage design leads to a remarkable efficiency. Apart from that, MULTIJECTORs® posses no revolving parts, require no lubrication nor maintenance and create no heat. The construction size is small which allows to apply the vacuum exactly where it is needed. This avoids long pipelines to transfer the vacuum which always involves energy losses. Moreover the available space above reactors or mixers in the chemical industry is limited. Applications in EX-areas (ATEX) are no problem, since the MULTIJECTOR® is free of ignition sources. There is no electricity necessary, the whole Vacuum Conveyors can be run and controlled pneumatically. Multiple-stage ejectors can be manufactured in different materials (stainless steel, plastic, etc.) or can be coated if, for example aggressive vapours are aspirated through the system.

At the typical discontinuous vacuum conveying the vacuum is switched on and off rapidly according to the suction and discharging cycles. This is easily done with MULTIJECTORs® by a valve on the energy supply side. During the necessary powder discharging cycle no energy is consumed. With mechanical/electrical pumps this operation mode is not possible. Here the vacuum pump must run continuously. In addition, the multiple-stage, compressed air-driven pump ensures an easy aspiration of the system since the vacuum connection is only separated from exhaust side by the gas jet, that is when the pump is running. Finally the performance of the jet pump can be indefinitely changed just by adjusting the feeding pressure (e.g. between 4 and 6 bar) to optimize energy consumption.

17.3.2
Conveying and Receiver Vessels

In chemical processing it is often required to convey a range of different powders and granules with the same vacuum conveyor. For example, in the paint industry various dyes and pigments are to be transferred. Therefore the design must be simple and should allow easy dismantling and cleaning.

Simultaneously the chosen material for the powder contact parts must be resistant against aggressive chemicals and cleaning liquids. These reasons lead to a modular stainless steel design which allows a quick change of the materials to be conveyed as well as it fulfils the strict hygienic demands (GMP, FDA, WIP, CIP) in chemical-pharmaceutical processes (Figure 17.10).

Separate single stainless-steel modules allow individual optimisations for the specific conveying tasks. A successful conveying may, for example just depend on the detail design of the suction module which could have a radial or tangential inlet (Figure 17.11). Tangential inlet is applied for extremely fine powders such as toner or soot and keeps the filter surface load low. The collection of the material in the lower half of the conveyor is improved if in addition a further cyclone cone is inserted.

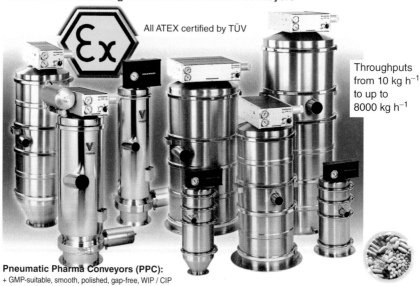

Receiver-and collecting containers for Vacuum Conveyors

All ATEX certified by TÜV

Throughputs from 10 kg h^{-1} to up to 8000 kg h^{-1}

Pneumatic Pharma Conveyors (PPC):
+ GMP-suitable, smooth, polished, gap-free, WIP / CIP

Figure 17.10 Conveying and receiver vessels in modular, single-body and pressure-rated design. (Volkmann Vakuum Technik GmbH.)

Modular container design:
1. Highest flexibility by modular construction system
2. Different flow conditions during suction

Figure 17.11 Different flow schemes with radial and tangential inlet modules. (Volkmann Vakuum Technik GmbH.)

However, when conveying wet filter cakes or other sticky and adherent powders the radial inlet gives better performance because here a different internal flow avoids baking and build up of powders at the container wall. The conveying and receiver vessels can be designed in order to avoid segregation and separation which is especially important if ready-made mixtures are conveyed. A typical application is the feeding of powder mixtures into tablet presses. Since the active ingredient in a pharmaceutical mixture may only be present in a very low proportion it is vital that this small percentage is uniformly spread over the whole mix. Independent research institutes and tablet manufacturers confirmed that MULTIJECTOR® Vacuum Conveying System is able to convey this mixture without segregation. The detail configuration of the individual powder transfer system is often found by trials.

In combination with the multiple-stage, compressed-air driven vacuum pumps the vacuum conveyors are small and lightweight which allows flexible mounting or mobile units. Due to the high achievable negative pressure it is also vacuum proof. Depending on the size of the receiver vessels and the power of the vacuum pump, a certain throughput capacity is reached. With volumetric transport, the performance depends on the bulk density. Figure 17.12 shows examples for three different vessels diameters: 170, 315 and 450 mm.

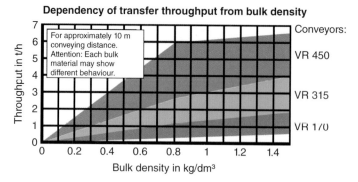

Figure 17.12 Capacity over bulk density for three different vacuum conveyors.

17.3.3
Filter Systems

Whilst entering the vacuum conveyor, the conveyed material is decelerated by the increased diameter of the receiver vessel and therefore the main bulk material is collected at the bottom section of the conveyor.

According to the particle size distribution a more-or-less small fine dust is moved towards the filter unit which is located at the top, below the vacuum source. The filter unit should achieve a high separation rate (dustless conveying) whilst maintaining the pressure difference low. With modern filters the exhaust air from the vacuum pump can be blown in the environment whereas with toxic, potent or cancerous powders further treatment is necessary. Precise demands are given by filter classes, for example from BIA [5] (German employer's liability insurance association) or other authorities (Table 17.1).

The extraordinary variety of applications for vacuum conveyors made it necessary to maintain the modular and flexible design even for the filter system. A filter change should be possible in a very short time without any tools. This is only possible in combination with the modular receiver vessels.

Basically the filtration systems can be divided in deep filtration and surface filtration. Surface filters do not separate the dusts in the depth of the filtration material (impacts, diffusion, electrostatical effects) but separate the particles at a surface of a membrane. However the surface coating (e.g. PTFE) can be sensitive against mechanical loads and rough cleaning. This has to be considered if many product changes occur. In general, solid filter materials and candles have a much longer lifetime than simple filter bags. Standard filter bags and their mounting systems have a limited separation factor and after a longer use or washing fibres may drop out of the cloth and contaminate the material to be conveyed (Figure 17.13) [6].

In chemical processes filter out of sintered polyethylene, polyester, PTFE, stainless steel and cellulose are used. With the standard cyclic operation the discharging time is used for an efficient filter cleaning with backblowing air shocks. Filter blockages can be avoided as well as maintenance.

Table 17.1 Different standard categories for solid filters.

NORM	Eurovent	DIN 24185	DIN	DIN	Test aerosol	Filtration grade	IEC	Average transmittance grade	Test aerosol	BIA	Test aerosol	Average transmittance grade
		DIN 24184	EN 779	EN 1822		[> %]	60335-2-69 Annex AA	[> %]				[> %]
COARSE FILTERS	EU 1	EU 1	G 1		Synthetic dust	<65	L	5	Quartz dust	U	Quartz dust	5
	EU 2	EU 2	G 2			65-80						
	EU 3	EU 3	G 3			80-90				S		1
	EU 4	EU 4	G 4			≥90						
FINE FILTERS	EU 5	EU 5	F 5		Atmospheric dust	40-60	M	0,1		G		0,5
	EU 6	EU 6	F 6			60-80						
	EU 7	EU 7	F 7			80-90				C		0,1
	EU 8	EU 8	F 8			90-95						
	EU 9	EU 9	F 9			≥95						
HEPA FILTERS (HEPA = High Efficiency Particulate Air Filter)	EU 10	Q		H 10	DEHS cl. DOP*	85	H	0,005	Paraffine oil mist	fine oil	Paraffine oil mist	0,05
	EU 11	R		H 11		95						
	EU 12	R		H 12		99,5						
	EU 13	S		H 13		99,95						
	EU 14	(T)		H 14		99,995						
ULPA FILTERS (ULPA = Ultra Low Penetration Air Filter)	EU 15	(U)		U 15		99,9995						
	EU 16	(U)		U 16		99,99995						
	EU 17	(U)		U 17		99,999995						
					Attention!		A direct reference from BIA- and IEC-classes to other class systems is not possible!					

*DEHS = Di-ethyl-hexyl-sebacat   DOP = Di-octyl-phtalat

17.4
Continuous Vacuum Conveying

The principle of discontinuous conveying allows a simple design but reaches boundaries when it comes to conveying heights of more than 30 m and distances of 100 m. With the combination of two receiver vessels and if the pneumatic control is switched to the twin mode a continuous vacuum in the conveying line is maintained (Figure 17.14). Alternatively a rotary valve can be used.

17.5
Reactor- and Stirring Vessel Loading in the Chemical Industry

The specific components of pneumatic MULTIJECTOR® Vacuum Conveyors allow the usage in all relevant dust and some gas EX zones. Details are described in the ATEX certificates published by the German TÜV.

Filter systems

* Deep filtration:

e.g. sintered candle filters, plastics, stainless steel, cellulose, bag filters, tube filters

* Surface filtration:

Membrane filters, e.g. PTFE coated filters

* Primary / secondary filters

Figure 17.13 Typical exchangeable filter units for vacuum conveying systems. (Volkmann Vakuum Technik GmbH.)

Continuous vacuum conveying

Figure 17.14 Continuous vacuum conveying; for example for the discharging of Big-Bags and sacks according to a given recipe and loading of a mixer in an EX area. (Volkmann Vakuum Technik GmbH.)

Powder locking, inerting with INEX vacuum conveyors

Figure 17.15 Feeding of solids in gas EX zones 0 and 1 with INEX. (Volkmann Vakuum Technik GmbH.)

In order to ensure safe loading of powders into gas EX zones 0 and 1 (hybrid mixtures), the MULTIJECTOR vacuum conveyors can be expanded into so called INEX systems (Figure 17.15). An integrated powder lock and inerting system eliminates any possible ignition source. The German TÜV has also published a further ATEX certificate for the INEX vacuum conveyors. With that, there is now a safe alternative available since the up-to-now used methods lead to many accidents.

Pressure rated INEX vacuum cnveyors

Figure 17.16 Feeding of powders in reactors with pressure-rated INEX conveyors. (Volkmann Vakuum Technik GmbH.)

Powder transfer, weighing and dosing in one step: CONWEIGH / VAWIDOS

Figure 17.17 Vacuum conveyor on a scale – powder transfer, weighing and dosing. (Volkmann Vakuum Technik GmbH.)

Powder feeding, Conveying and Gravimetric dosing station

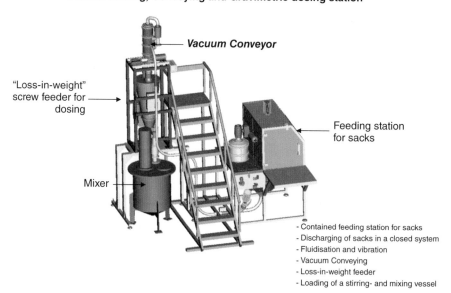

- Contained feeding station for sacks
- Discharging of sacks in a closed system
- Fluidisation and vibration
- Vacuum Conveying
- Loss-in-weight feeder
- Loading of a stirring- and mixing vessel

Figure 17.18 Vacuum conveyor integrated in a feeding process, dosing and mixer loading. (Volkmann Vakuum Technik GmbH.)

17.5 Reactor- and Stirring Vessel Loading in the Chemical Industry | 329

Table 17.2 Short extraction out of the Volkmann GmbH conveying parameter database.

Application examples

Material to be conveyed	Bulk density (kg dm^{-3})	Conveying height (m)	Conveying distance (m)	Particle size	Material characteristic	Transporting capacity (kg h^{-1})	Suction- and loading system
Dicyano-diamide powder	0.7	2	4	20 up to 100 μm	Bridging occurs	600	VR 315: Weight = 19 kg Ø 345 mm, Height = 955 mm
Activated carbon dust	0.4	3	3	Fine dust, 5 up to 80 μm	Adherent	120	VR 170: Weight = 6 kg Ø 180 mm, Height = 450 mm
Placebo granule	0.6	2	4	100 up to 1000 μm	Very good flowing	5.000	VR 450: Weight = 39 kg Ø 480 mm, H = 1.160 mm
Hexamethyl-enetetramine	0.7–0.8	4	12	>5 μm	Hygroscopic	1.400	VR 315: Weight = 19 kg Ø 345 mm, Height = 955 mm
Paracetamol powder	0.8	1.5	3	40 up to 200 μm	Adherent and bridging	1.000	VR 315: W = 19 kg Ø 345 mm, H = 955 mm
Ferric powder	3.5	4	5	50 up to 100 μm	Sensitive, abrasive	180	VR 170: W = 6 kg Ø 180 mm, H = 450 mm
Flame soot	0.2	5	15	0.1 up to 44 μm	Bridging, adherent	1.440	VR 450 Contivac (continuous conveying)
Titanium-dioxide and dyes	0.85–1.4	1.5	3	>2 μm	Partial strongly adherent	600 up to 1.628	VR 315: W = 19 kg Ø 345 mm, H = 955 mm
Shredded wheat pellets	0.98	5	12	Dust content, Ø 2 × 8 mm	Sensitive	3.600	VR 450: W = 39 kg Ø 480 mm, H = 1.160 mm
Sand-lime-mixture	1.38	0,5	32	400 up to 900 μm	Bad flowing	30.0	VR 170: W = 6 kg Ø 180 mm, H = 450 mm
Toner	0.2–0.8	4	5	Very fine dust >0.1 μm	Aggressive	1.100	VR 315: W = 19 kg Ø 345 mm, H = 955 mm
Polyurethane granule	0.9	10	12	1 up to 3 mm	Good flowing	1.000	VR 315: W=19 kg Ø 345 mm, H=955 mm
Plastic caps	—	0.5	10	Ø 40 mm, H = 25 mm	Sensitive	17.100 caps per h	VR 315: W = 19 kg Ø 345 mm, H = 955 mm

If the solids are to be loaded in a reactor which is already under positive pressure, INEX Vacuum Conveyors can be manufactured in a pressure-rated design (Figure 17.16).

17.6
Conveying, Weighing, Dosing and Big-Bag Filling and Discharging

MULTIJECTOR® vacuum conveyors for powders and other bulk materials can be combined with many other processes. Here Figures 17.17 and 17.18 should only give some ideas about the many different possibilities.

17.7
Application Parameters

As could be seen in Chapter 1, countless parameters may affect the behaviour of the materials to be conveyed. As a conclusion, Table 17.2 shows the parameters with the strongest influence and some examples are given. This represents just a short extraction from a huge data base held by the author. These figures were all found with the aid of 1 : 1 scale vacuum conveying trials done in Soest, Germany. It can be said that there are only very rare cases known where a vacuum conveyor might not be able to improve the bulk material handling in chemical, pharmaceutical or food processing.

References

1. Bohl, W. (1989) *Technische Strömungslehre*, 8. Auflage, Vogel Buchverlag, Würzburg.
2. Kleffmann, O. (1991/1992) *Strömungsmechanik, Vorlesungsunterlagen*, Universität-GH-Paderborn, Abteilung Soest, Soest.
3. Spurk, J.H. (1996) *Strömungslehre: Einführung in die Theorie der Strömungen*, 4. Auflage, Springer-Verlag, Berlin Heidelberg.
4. Siegel, W. (1991) *Pneumatische Förderung: Grundlagen, Auslegung, Anlagenbau, Betrieb*, 1. Auflage, Vogel Buchverlag, Würzburg.
5. Berufsgenossenschaftliches Institut für Arbeitssicherheit (BIA) *Vergleichende Darstellung der Klassifizierungen von Staubbeseitigenden Maschinen*, ZH1/487, Hauptverband der gewerblichen Berufsgenossenschaften e.V., Sankt Augustin.
6. Ramme, T. (1997) *Saugförderung von Pulvern, Stäuben, Granulaten, und Kleinteilen*, Fachbeitrag der Fa. VOLKMANN in Schüttgut Nr.2, Trans Tech Publications, Clausthal-Zellerfeld.

18
Vacuum Filtration – System and Equipment Technology, Range and Examples of Applications, Designs

Franz Tomasko

18.1
Vacuum Filtration, a Mechanical Separation Process

Alongside chemical and thermal processes for the separation of material mixtures in process engineering, there are also the mechanical processes, which are counted amongst the physical separation techniques. Within these separation techniques the components of heterogeneous material mixtures are separated in such a way that their aggregate condition and their chemical composition remain unaltered. Heterogeneous material mixtures of this type, as a rule polydisperse systems or slurries (consisting of solid and liquid phase) are separated by mechanical methods such as screening, upward current classification, sedimentation, filtration and so on.

The mechanical separation process most frequently used is filtration. This is always carried out under the influence of a pressure difference. We talk of filtration under 'liquid pressure' and under 'gas pressure' and thus differentiate between pressure and vacuum filtration. In both cases, the liquid phase is forced (or sucked) by the pressure difference through a filter medium, while the solid phase remains on the filter medium.

If the liquid pressure works through the slurry itself, either due to the force of gravity or by pumping, this is known as pressure filtration and filters which work on this principle are known as pressure filters. According to requirements, very high pressures may be selected.

In filtration under gas pressure, a vacuum is created whereby the liquid is sucked through the filter medium while the solids remain on the surface of the filter. In theory the maximum gas pressure created would be an absolute vacuum. As a rule, liquid ring vacuum pumps are used to create the vacuum and filters which operate based upon the principle of vacuum filtration are known as 'vacuum filters'.

In addition, there are special cases in which filters are operated using gas pressure.

In the following, vacuum filtration will be presented as a continuous filtration process, as this feature of a continuously operable filter is of a distinct significance for vacuum filtration, and it is therefore often preferred over pressure filtration which can only be operated on a discontinuous basis. Apart from this, the following chapters do address filtration systems in which filter-cakes are formed, meaning that a filter cake is formed on the filter medium (most commonly a filter cloth), which then itself works as a filter medium.

18.1.1
On the Theory of Filtration and Significance of the Laboratory Experiment

Filtration as a mechanical solid–liquid separation process is a basic procedure in process engineering, and theoretically it is nevertheless relatively seldom mastered. One of the reasons for this is that surface boundary forces play an outstandingly important role in this process, and until recently they have to a large extent denied the scientific community access to their secrets.

In spite of this, there are today computer programmes [1–3], which are gaining in significance for operational practice, based on the Hagen–Poiseuille Law which has been known for decades, and the experimental equations of Darcy, Kozeny–Carman [4].

All these systems and their basic equations, however, start from a base of idealised, homogeneous model representations (circular capillaries of the same lengths, heaps of pebbles of a defined diameter).

In practice the slurries to be filtered are of heterogeneous nature and would require the determination of always new parameters in order to be in a position to calculate the industrial filter size.

Since capturing of these parameters is extremely time-consuming and, despite that, always less effective than an empirical filtration experiment in the laboratory or pilot plant, the latter is in practice normally preferred as the basis for designs [5], or integrated as an indispensable component in the computer programmes described above [6]. Figure 18.1 illustrates the schematic construction of an experimental arrangement for lab-scale (test leaf) experiments.

One of the expert systems/computational tools for sizing worthwhile highlighting is the filter sizing software – called FILOS – which has been on the market for some years now. It is regularly optimised and it combines the available theoretical basics with a manageable amount of laboratory tests for determination of the key parameters which characterise the filtration process. This system is a very useful additional tool to the conventional lab-scale tests for filter sizing [7].

Essentially, the experimental apparatus illustrated in Figure 18.1 consists of a test leaf, a vacuum generator, a stirring rod, glass containers for holding the slurry, the wash liquid and the filtrate, connecting hoses and fittings. A stopwatch, a thermometer, a measuring cylinder and a set of scales complete the equipment.

By simulating all stages of the process which will later affect the filtration, both in chronological order and duration as well as through observation of the direction of filtration during the respective cycle phase, results are obtained which provide by graphical analysis a clear picture of the corresponding interdependencies in the production process.

18.1 Vacuum Filtration, a Mechanical Separation Process

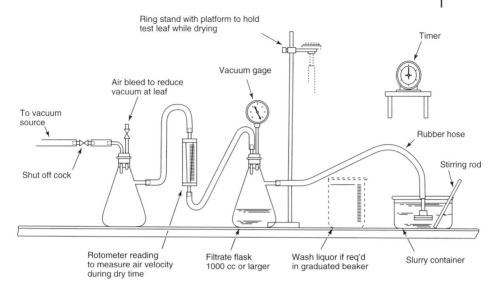

Figure 18.1 Schematic construction of an experimental arrangement for lab-scale experiments.

Figures 18.2 and 18.3 show the typical interdependencies of the parameters specific filtration capacity, filter-cake residual moisture and cycle time.

Values regarding specific filtration capacity, residual moisture in the filter cake and washing efficiency (if washing of the filter cake is necessary) as well as their dependency on cycle time and immersion depth established in the laboratory can, within the limitations set by the operational aggregate and taking into account of safety factors, reliably be applied to the production unit. The laboratory experiment also provides important information regarding the most appropriate method of discharging the filter cake, and the filter cloth to be selected.

Attention should also be drawn here to the further possibility of obtaining relevant information concerning the long-term behaviour of the respective filter aggregate by means of pilot plant operations.

18.1.2
Guide to Filter-Type Selection

Filtration on a continuously operating filter is a series of individual operations, essentially made up by formation of the filter cake, filtrate draining and discharge of the filter cake. The factors which control the course of every individual operation are thus also decisive for the selection of filter type.

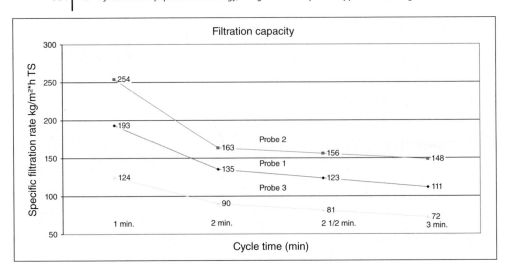

Figure 18.2 Specific filtration capacity in kg m^{-2}·h^{-1} dry solids as a function of the cycle time.

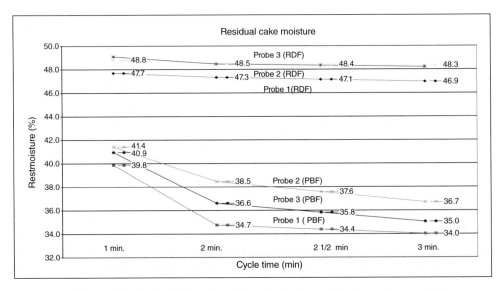

Figure 18.3 Residual filter-cake moisture in % w/w as a function of the cycle time.

The most important factors influencing the selection of the filter type include particle size distribution in the solids to be separated, the specific gravity of these solids and the solids content of the slurry to be filtered.

The solids content of the slurry is particularly relevant for the achievable filtration capacity. Here the rule of thumb is that slurries with a solids content of <20 g l^{-1} or those which in a cake-forming time of 5 min fail to achieve a cake thickness of at least 3 mm are unsuitable for traditional vacuum filtration. In such

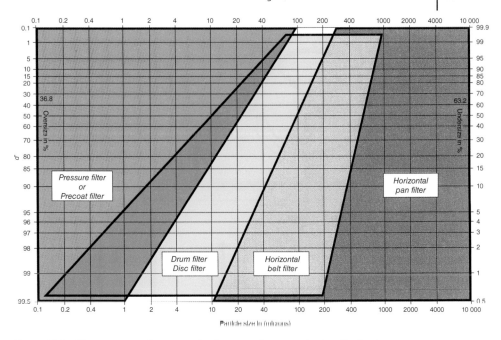

Figure 18.4 Diagram presenting type selection according to grain-size distribution.

cases, which almost invariably occur when handling dispersed fine sludges, it is necessary to turn to using special vacuum drum filters, the so-called precoat filter, or to liquid pressure filters.

However, where there is a coarser particle size distribution, this more often goes hand-in-hand with a higher concentration of solids in the slurry to be filtered. In this case, one can assume that continuous vacuum filtration will be successful. The particle size distribution curve of the solids to be separated is thus a suitable indicator for the selection of filter type [8].

Figure 18.4 is a diagram representing the frequently used filter types in relation to particle size distribution. The overlap areas shown here are to be understood as zones with flowing boundaries for which the most cost-effective design must be specified subsequently by the use of the experiments outlined above. In doing so, further factors such as throughput, residual moisture, washing efficiency (where necessary) as well as the type of discharge of the filter cake must also be taken into consideration.

18.2
Design of an Industrial Vacuum Filter Station

Vacuum filtration plants consist not merely of the respective vacuum filter, but much more of a series of complementary units which only make up the effective

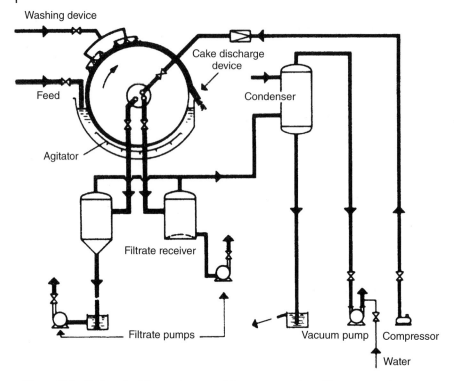

Figure 18.5 Design of a filter station comprising a vacuum drum filter as example.

production system when they are assembled. Figure 18.5 shows the typical design of a filter station using the example of a vacuum drum filter.

The pressure difference required for vacuum filtration is generated for this purpose (as in most industrial applications) economically by the use of a liquid ring vacuum pump. The amount of vacuum applied depends on many parameters, not least on the porosity of the filter cake. As a rule of thumb as vacuum requirement one normally assumes about 0.5 m^3 min^{-1} per square metre filter area at about 200–400 mbar absolute at the vacuum pump suction connection.

The filtrate recovered is separated by one filtrate receiver each from the liquid–gas mixture. The filtrate can either be discharged by a suitable pump or via a barometric leg. If vapour pressure, temperature and chemical properties of the slurry are such that the occurrence of undesirable vapours is likely, the use of a condenser is to be recommended. This is installed between separator and vacuum generator.

For discharge of the filter cake from the drum by a scraper, a pulsating compressed air feed is provided. The amounts and the pressure required for this in most cases allow for a feed to be taken from the factory supply.

The rule of thumb for compressed air consumption is about 0.2 m^3 min^{-1} per square metre filter area at approximately 500 mbar pressure. If there is no factory supply, then a rotary compressor will normally fulfil the task.

Slurry feed is from the plant pipework directly into the filter trough. Washing liquid is applied with suitable equipment to the filter cake which was formed on the drum. To regulate the level in the filter trough, an adjustable overflow connection or a commercially available level control can be provided.

18.3 Methods of Continuous Vacuum Filtration, Types of Design and Examples of Application

The principle of vacuum filtration is implemented on a large number of industrial units of different designs. The broad enormous range of available designs makes it essential to restrict the subject to a certain extent.

This article is concerned solely with the field of continuous vacuum filtration and differentiates, according to the direction of filtration and the geometrical shape of the filter area, between four principles:

- vacuum filtration on a curved convex surface,
- vacuum filtration on a curved concave surface,
- vacuum filtration on a flat horizontal surface,
- vacuum filtration on a flat vertical surface.

The following discourse is oriented towards these criteria and highlights for each of the types of design typical fields and examples of application.

18.3.1 Vacuum Filtration on a Curved Convex Surface, the Drum Filter

The principle of continuous vacuum filtration on a curved convex surface, with filtration in the opposite direction to the force of gravity is applied in the most common type of filter, the vacuum drum filters.

The standard drum filter is particularly suitable for continuous processes and due to its special adaptability offers a ready solution to the most widely varied of filtration problems; it is invariably suitable when dealing with sludges with a solid content of over 5% w/w and solids that can be kept in suspension by an agitator oscillating in the filter trough.

It can also be used without difficulty for intermittent operation, since the filling volume of the trough is limited and all the filter components are easily accessible for cleaning.

18.3.1.1 Design of a Vacuum Drum Filter

The main components of a drum filter are the filter drum, the filter trough, the control valve, the agitator and the device for discharge of the filter cake.

Figure 18.6 is a schematic illustration of the design of a vacuum drum filter with external filtrate pipework.

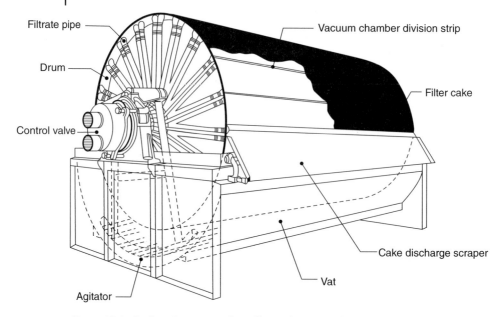

Figure 18.6 Design of a vacuum drum filter with external filtrate pipework.

The Filter Drum The filter drum is an enclosed cylinder lying horizontally, suspended to rotate around its centre axis. The interior is reinforced by means of intermediate rings. The filtration area on the drum surface is divided into independent cells by division strips having a swallowtail-shaped groove. At the head ends of the drum there are also grooved strips applied.

The drum deck is formed by replaceable filter grids (sieve-plates), which support the filter cloth required for retention of the solids. The filter cloth (usually fabric or metal) is placed into the division strip grooves and fixed by means of a rubber caulk or similar. The supporting filter grids are honeycomb in form, whereby the largest possible cross-section and high filtration capacities are achieved. In the propylene version, as a result of the non-sticky properties of this material, encrustation is virtually excluded. At one end of the drum, there are radial outlet connections for each cell for the filtrate draw off. These are readily accessible, and do not project beyond the level of the drum deck. For leading the filtrate away, the connections are joined to filtrate pipes by means of sleeves.

The filter cloth required for retention of the solids is fixed into the grooves by means of rubber strings or similar.

The Control Valve The other end of the filtrate pipe discharges after a 90° bend into the control valve faceplate, arranged parallel to the face of the drum. The so-called control plate is fixed to the control valve and forms the counterpiece to the faceplate. Various filtrate pipe sections are connected through apertures in the control plate. These apertures control the filter cycle inasmuch as, according to its position, the cell is evacuated by the vacuum connection, blocked or blown out with compressed air. For connection with the vacuum system, comprising filtrate separator(s) (possibly also condenser) and vacuum pump, the control valve is normally fitted with two nozzles.

The Filter Trough The filter trough is charged with the slurry to be filtered and at the same time acts as the foundation for the filter drum bearings and the remaining components of the filter. The level to which the trough can be filled is adjustable.

The Agitator The agitator prevents the slurry from segregation and the solids from sedimentation. It is actuated by an electric motor with a slip-on gear and oscillated by an eccentric.

The Drum Drive The drum drive is mounted on the side of the trough opposite the frequency invertor control valve. The driving unit consists of an electric motor, regulator and slip-on gear and for a drum rotation speed adjustable over a wide range.

Frequency-controlled drives are state of the art.

Device for Discharge of the Filter Cake The device for discharge of the filter cake is arranged on the immersion side of the filter drum on the long side of the trough. The discharge of the filter cake has a major influence on the function of a vacuum drum filter. For this reason, it will be examined closely below.

According to the properties of the slurry, the cake formed on the drum, washed and dried, displays irregular layer thicknesses, structure, consistency, adhesive behaviour and so on. While thick, dense, fibrous cakes can be blown and/or scraped off with no problem, discharge in the case of thinner cakes of a pasty consistency is much more difficult. Special arrangements are also necessary for discharge of the filter cakes in application with solids which tend to block the filter cloth. In these cases, constantly high filtration can only be achieved by using an effective method of filter cloth cleaning, which makes a special cake-discharge device an absolute necessity.

In practice, there are five essentially different types of discharge devices [9]. These are by no means alternatives, for each application, there is but one optimum method of filter-cake discharge.

The Scraper Discharge The discharge of the filter cake with scraper and short blow back with compressed air is the simplest method. With this device, a blade is installed tangentially to the filter drum, without affecting however the filter cloth and/or the wire winding required additionally for this type of discharge while the filter drum is rotating (Figure 18.7).

The scraper is suspended at both ends in rotating bearings and has an adjustment to enable it to be fixed at a minimum spacing from the drum. The blade of the scraper is normally clipped on a plastic blade.

After blocking the vacuum from the filter cell arriving at the discharge point there is a short blow back of compressed air which expands the filter cloth between

 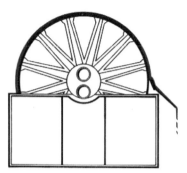

Figure 18.7 Scraper discharge.

the wire winding fitted at a spacing of 30 to 50 mm. This loosens the filter cake from the cloth Subsequently, the cake is discharged by the scraper. At the same time, the filter cloth is blown free from the rear by of compressed air.

The scraper discharge method can be applied in almost all filtration problems in which the filter cake reaches a thickness on the drum of at least 3 mm.

The Roller Discharge With this discharge device, a roller rotating counter to the direction of the drum carries out the discharge of the filter cake from the drum. Here, the speed at the circumference of the roller is slightly higher than that of the drum. The roller is either driven by the drum drive or it has its own frequency-controlled drive. The discharge of the cake from the roller is carried out by a scraper or comb. A residual layer of the cake remains on the roller, causing better adhesion to the filter cake and consequently an easy transfer (Figure 18.8).

 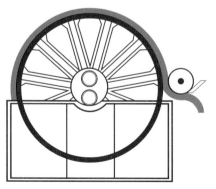

Figure 18.8 Roller discharge.

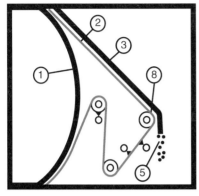 

1: Filter drum
2: Filter cloth
3: Filter cake
5: Filter cake discharged
8: Reversing roller

Figure 18.9 Cloth discharge.

Filters with roller discharge do not require a wire winding. They are used in cases where the fine suspended particles, for example pigments (which can only form a cake of 0.5–3 mm in thickness on the filter drum) have to be filtered. The possibility of using this discharge method has emerged with the development of monofilament plastic weaves, as despite the fineness of the solids these do not tend to lead to blinding.

The Cloth Discharge The cloth discharge system provides an answer to many filtration problems where continuous filtration cannot normally be used, for example because of the fineness of the suspended particles (Figure 18.9).

A fully automatic reversing roller system arranged in parallel to the drum axis is used to remove the cloth. Usually, the system consists of a roller to remove the cake, a reversal roller and a tensioning roller. The cake removing roller loosens the cake from the filter cloth. In the region of the reversal roller, the filter cloth can, when necessary, be washed free of cake from both sides and regenerated once more for the next filtration. In case of adhesion to the cloth, discharge of the cake is assisted by a scraper or similar. Parallel running of the rotating filter cloth is optimally controlled so that there are no major track deviations ($\pm$ 5 mm). For controlling the cloth, there is a choice between three fully automatic systems:

1) *Mechanical cloth control*: In this system, the edges of the filter cloth have guides of plastic or a similar material attached. When the edges of the cloth are discharged from the drum, they are directed into tracks so that there is no appreciable deviation from the path before the cloth is fed back onto the filter drum.

2) *Electromechanical cloth control*: In this system there is a sensor which detects the edge of the cloth; if there is a sideways deviation, it sends an impulse to an adjustment mechanism. This will then, according to the impulse, adjust the control roller (swivel roller) in its parallel position to the drum and direct the cloth back onto its normal or central course.

3) *Pneumatic cloth control*: In this system the movement of the control roller (swivel roller) is activated by a sensor paddle and a solenoid valve controlling the flow of compressed air into one of two bellows (one bellow on each roller end), which inflates to move the roller in one or the other direction, which aligns the cloth by moving one end of the roller forward or backwards to track the cloth back to its centred position on the filter.

The Chain or String Discharge   In this system of discharge, chains or strings run along the filter cloth and in the discharge zone are led away from the drum, then reversed and led back to the drum (Figure 18.10). A guiding comb ensures uniform spacing of the chains or strings. This discharge device is used when very thick filter cakes are formed such as in the separation of coarse crystalline material. Chain or string discharge is also a viable option in the case of fleecy or fibrous filter cakes.

The Precoat Discharge   This is a special type of discharge of the filter cake used on the so-called precoat filters. Precoat filters are drum filters in which the filter medium is composed of a layer of diatomite, perlite, wood-flour, synthetic granulate, and so on. This layer is applied in thicknesses of approximately 100 mm before the actual filtration of slurries mostly having a very low solids content and with very fine particulate composition (Figure 18.11) commences.

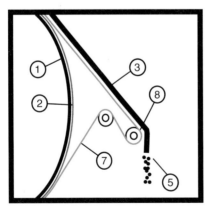

1: Filter drum
2: Filter cloth
3: Filter cake
5: Discharged cake
7: Strings
8: Reversing roller

Figure 18.10   String discharge.

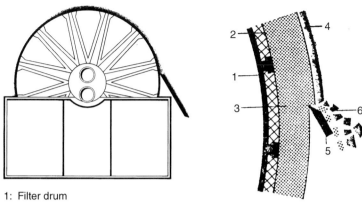

1: Filter drum
2: Filter cloth
3: Precoat coating
4: Filter cake
5: Scraper
6: Cake discharge

Figure 18.11 Precoat discharge.

The solids deposited during filtration blind the top surface of the process layer, which consequently has to be continuously regenerated. This is carried out through continuous scraping off from the top surface of the filter cake layer.

Precoat discharges of modern design are operated with a separate drive for scraper advance and with a second drive for the scraper fast return motion. The advance speed is thus independent of the drum speed and continuously adjustable, within the framework of limits applying to the respective use (e.g. from 0.1 to 0.002 mm min^{-1}). The scraper advance is automatically shut off as soon as the preset minimum layer thickness is reached. In older designs, the scraper advance is mechanically regulated by rotation of the drum and operated in steps from 1/100 mm up to some tenths of a millimetre per rotation.

Precoat filters are outstandingly suitable for so-called polishing filtration: clearing pectin solutions, fruit juice filtration after fermentation and clarification, citric acid separation from mycelium-mash, dextrose cleaning, treatment of toxic industrial waste water, cleaning galvanic waste water and so on.

Additional Devices To account for certain process conditions, vacuum drum filters can be equipped with a variety of accessories. These include

- cake-washing devices (wash pipes or wash boxes),
- wash belt for supporting the filter-cake washing process,
- filter-cake pressing devices such as press rollers, press belts or so-called flapper rollers for further removal of moisture from the filter cake,
- covering hoods or vapour hoods, hoods for gastight enclosures,
- wire winding device for automatically winding the filter cloth onto the filter drum and fixing it with wire.

Figure 18.12 Dewaxing filter with zero solvent loss and zero emission (with gas tight hood).

Figure 18.13 Parts in contact with the slurry in rubber-lined mild steel design and with two cake-wash pipes on top of the filter, filtration area 25 m^2.

Well-thought-out use of these additional devices can guarantee optimum working performance of the filter.

The following Figures 18.12 and 18.13 show two examples for the aforementioned additional devices.

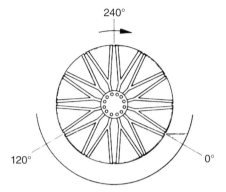

Figure 18.14 Working method of a drum filter with filtrate pipes installed externally.

18.3.1.2 Working Method of a Continuous Operating Vacuum Drum Filter

The filter deck of a vacuum drum filter consists – as already described – of a number of independent filter cells which we may be regarded – in historical development terms – as separate suction filters.

Each of the cells goes under all the phases in sequence, according to the process in the discontinuous suction filter, from formation of the filter cake, through washing and drying to discharge of the filter cake. Through the definition of filter cycle on the control plate, the chronological sequence of the cycle is assured and continuous operation of the vacuum filter is realised.

Figure 18.14 shows a filter drum which immerses from 0° to 120° in the slurry. We talk in this case of an immersion depth of 33 1/3%, related to the circumference of the drum (filter area). From approximately 30°, the cell is evacuated. This causes a pressure differential between the cell and the surface of the drum, that is the filter cloth surface. Due to the external atmospheric pressure, liquid is forced through the filter cloth. The solids in the slurry are retained by the cloth, and thus the formation of the filter cake starts.

As the series of cells run through the slurry, the filter cake continues to grow in thickness, depending on the porosity of the filter cake. The reason for the evacuation not starting until 30° lies in the fact that the smallest particles collect in the upper layers of the slurry in the trough despite the stirring by the agitator. If these fine solids are first to reach the filter cloth, the cake would become blind in texture and would offer great resistance to further liquid permeation. Filter performance would in that case be poor.

By delaying the start of filtration to about 30° it is ensured that there is a granular build-up of the filter cake starting with larger spaces between the particles and thus a greater drainage effect sets in. The fine particles finally form the surface of the cake.

After emergence of the cell, air passes through the filter cake. From then on, a two-phase mixture of liquid and air passes through the filtrate system. The air sucked through accelerates the rate of flow and ensures a good drainage of filtrate.

At the same time, air pulls liquid out of the spaces between the grains in the filter cake, meaning that the cake is dried.

As the cells continue rotating, if desired at between about 210° and 270°, the cake can be washed by adding liquid. This process is a combination of displacement washing and dilution washing. From 270° to 330°, the cake is again dried and then – after interruption of the vacuum – discharged from the drum by the filter-cake discharge mechanism.

Figure 18.15 illustrates the various possibilities of connecting the filtrate pipes to the cells.

Generally, there are two filtrate pipes per cell, one leading pipe at the start of each cell, seen from the direction of rotation, and one trailing pipe at the end of the cell. The pipes are joined in the shape of a Y hosepipe. This makes an immediate discharge of the filtrate possible, regardless of the actual location of the filter cell. This is the design to select particularly when it is required to separate mother- and wash-liquor.

For pure dewatering, that the without cake washing, filters are fitted with leading filtrate pipes only. Beyond top dead centre, this prevents air entering into the trailing filtrate pipe, which would be then always open, and thus not accelerating the filtrate in the leading pipe. When filters with heel cake are employed, it is appropriate to utilise trailing filtrate pipes only.

Each of the systems described is equally applicable for filters with external pipes and for those with internal pipework (Figure 18.16), which are used when there are specific filtration problems to be solved or when the filter areas are extremely large.

18.3.1.3 Different Constructions

According to the corrosion properties of the slurry, filters are most commonly constructed in carbon steel, rubber-lined or stainless-steel qualities.

For special parts, preference is given to using plastic, for the filter grids and the filtrate pipes, for example polypropylene. The control and near plates are normally made of high-molecular polyethylene. These parts can optionally also be supplied in other materials, for example steel, stainless steel or cast iron.

Parts which do not come into contact with the product are made of steel and may also have an appropriate corrosion-resistant paint coating. For light-duty applications, vacuum drum filters with main components made of plastic are also used.

18.3.1.4 Special Vacuum Drum Filters

Among others, these include those already mentioned above, viz. the precoat filter [11, 12], dewaxing filter [13], single-cell high-capacity vacuum filter [14], compact filter, pressure-belt drum filter and so on.

The compact filter is a skid type mobile apparatus in which the vacuum filter already has all the accessories connected and is thus ready to plug in for instant operation. There are versions available as single-cell filters or as cell filters having all the usual discharge arrangements, in addition to pre-coat filters with hydraulical lowering filter trough [15].

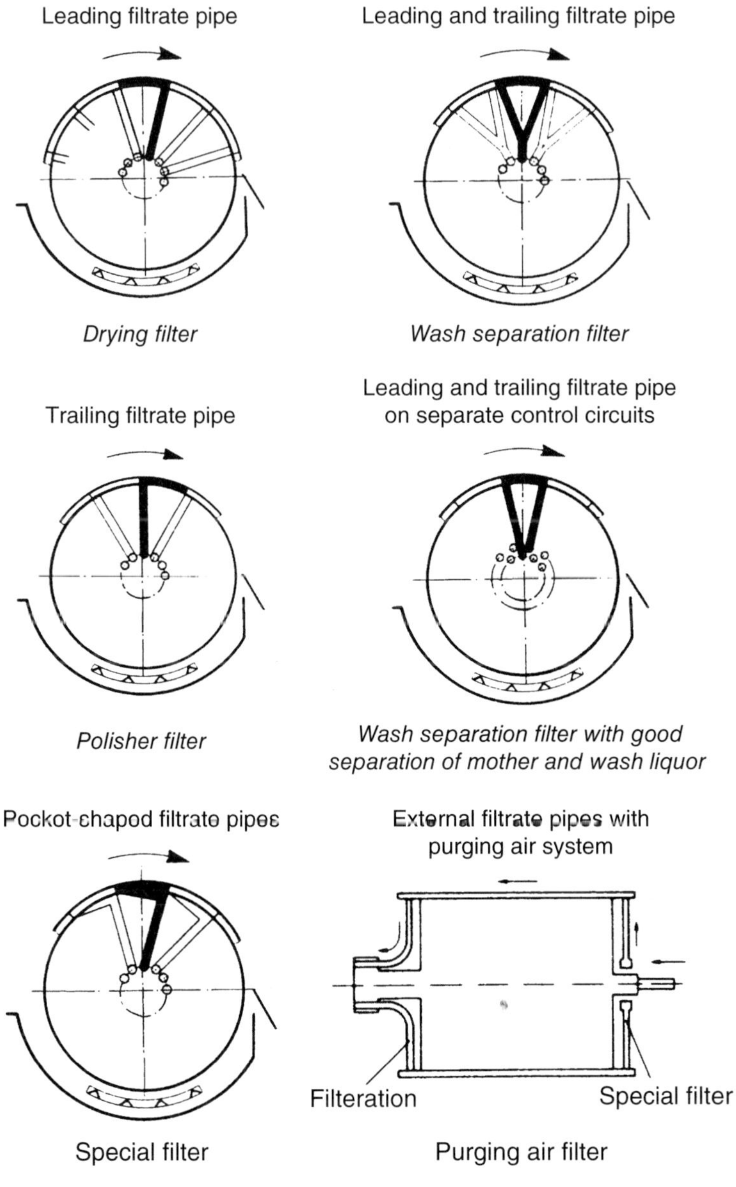

18.3 Methods of Continuous Vacuum Filtration, Types of Design and Examples of Application

Leading filtrate pipe

Drying filter

Trailing filtrate pipe

Polisher filter

Pocket-shaped filtrate pipes

Special filter

Leading and trailing filtrate pipe

Wash separation filter

Leading and trailing filtrate pipe on separate control circuits

Wash separation filter with good separation of mother and wash liquor

External filtrate pipes with purging air system

Filteration   Special filter

Purging air filter

Figure 18.15 Filtrate pipe systems [10].

The press belt drum filter [16] may be defined as a combination of a vacuum drum filter and a pressure stage. It comprises all the standard components of a traditional vacuum drum filter and a press belt attachment mounted on top. The standard components are of reinforced construction to withstand the additional stress imposed by the press belt equipment (up to 25 t for a 45-m^2 filter). This continuously operating special filter achieves a medium-specific reduction of the

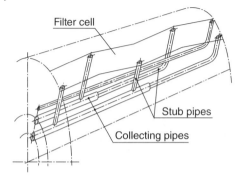

Figure 18.16 Filtrate pipe systems. Zelle = Filter cell; Stichrohre = Stub pipes; Sammelrohre = Collecting pipes.

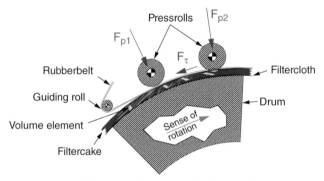

Figure 18.17 Principle sketch of the press belt function.

residual moisture level in the filter cake of up to 15 percentage points compared with the figures which can be attained by a traditional vacuum drum filter working under identical conditions. Figure 18.17 shows how the press rollers are arranged on the press belt drum filter and how the press belt device works, Figure 18.18 shows a press belt drum filter with a filtration area of 45 m² in rubber-lined mild steel design for the wetted parts, while Figure 18.19 shows a press belt drum filter with a filtration area of 20 m² including plastic hood constructed for the application in ATEX zone 0. Figure 18.20 gives details of tested press belt drum filter applications.

18.3.1.5 Calculation Example

It is assumed that on a conventional vacuum drum filter a suspension of any kind is dewatered to a residual moisture of 45% by weight in the filter cake discharged.

Through additional application of mechanical pressure this residual moisture is reduced to 30% w/w, that is 15 percentage – points less.

18.3 Methods of Continuous Vacuum Filtration, Types of Design and Examples of Application

Figure 18.18 Press belt drum filter for TiO_2 filtration, active filtration area $45\,m^2$.

Figure 18.19 Press belt drum filter with plastic hood for ATEX zone 0, active filtration area $20\,m^2$.

Disperse phase in feed slurry	Specific filtration rate kg m^{-2} h^{-1}	Vacuum filtration		Press belt filtration		Reduction in residual moisture		Fuel saving in downstream dryer
		Dry solids	Residual moisture	Dry solids	Residual moisture			
		%	%	%	%	%	%	%
Acrylic plastics	150	30	70	50	50	20	28.5	57
BaCO$_3$	150	65	35	82	18	17	48	59
Fly-ash from oxygen steel plant	130	64	36	80	20	16	44	55
CaCO$_3$	170	55	45	70	30	15	33	48
CaCO$_3$ precipitated	35	45	55	60	40	15	27	45
CaCO$_4$	400	65	35	75	25	10	28	38
Fe$_2$O$_3$	60	66	34	75	25	9	26	45
Kaolin	80	55	45	70	30	15	33	48
Maize gluten	12	40	60	48	52	8	13	28
Mg (OH)$_2$	200	51	49	62	38	11	22	36
MgCO$_3$	100	50	50	60	40	10	20	33
PbS-concentrate	1000	86	14	92.5	7.5	6.5	46	71
TiO$_2$ (hydrate)	60	38	62	45	55	7	13	25
TiO$_2$ (pigment)	80	54	46	62	38	8	17	28
Zeolite	130	45	55	52	48	7	13	25
ZnS-concentrate	900	83	17	89.5	10.5	6.5	38	60

Figure 18.20 Examples of press belt drum filter applications.

The theoretical energy savings induced hereby amount to

$$S = \left[1 - \left(\frac{Yf - \Delta}{Yp + \Delta}\right) \cdot \frac{Yp}{Yf}\right] \cdot 100\% \quad S = \left[1 - \left(\frac{30}{70}\right) \cdot \frac{55}{45}\right] \cdot 100\%$$

$$S = \left[1 - \left(\frac{45 - 15}{55 + 15}\right) \cdot \frac{55}{45}\right] \cdot 100\% \quad S = 48\%$$

Yf	=	weight % liquid
Yp	=	weight % solids
Δ	=	% points-moisture reduction
S	=	% energy savings.

Figure 18.21 Internal filter.

Even if the residual moisture is only reduced by 5% points, for example from 35% residual moisture to 30%, the energy savings yet amount to over 20%.

18.3.2
Vacuum Filtration on a Curved Concave Surface, the Internal Filter

The principle of continuous vacuum filtration on a curved concave surface with filtration in the direction of the force of gravity is realised using the internal filter. This is a vacuum drum filter, without trough and without agitator. Its simple and robust design is the secret behind its economy in operation.

The filtration process is essentially the same as on drum filters. The filter area however, is formed by the concave curve of the interior of the filter drum. The filter bed is also divided into several cells, which are connected by filtrate pipes with the control valve on the closed head side of the filter.

The slurry is fed in the direction of the force of gravity onto the concave-curved interior surface of the filter drum. On the basis of this method, the interior filter is particularly suited to the filtration of relatively fast-settling slurries, such as ore concentrates, sludge containing metal powder, and so on.

The filter cake held in place by the vacuum is lifted to the top of the drum's travel and falls off when the vacuum is shut off, or as the result of an additional blow back of air. The filter cake can then be discharged either by a simple chute or with the help of a belt conveyor. The drum is driven by means of a friction wheel. Figure 18.21 shows internal filters made by the former Dorr-Oliver GmbH nowadays FLSmidth Weisbaden, Germany.

18.3.3
Vacuum Filtration on a Flat Horizontal Surface

The principle of continuous vacuum filtration on a flat horizontal surface with filtration in the direction of gravity is utilised amongst others by belt filters and pan filters.

18.3.3.1 The Belt Filter

The slurry is fed from above by a variety of feed-in systems onto a filter surface which is arranged horizontally. On here is where the whole filter cycle takes place, and where a number of process and technical engineering problems are more beneficially solved than is the case with the drum filter. This led, particularly after WWII, to a rapid expansion of use of the belt filter [17]. Typical applications today include the chemical industry (zeolite, various salts, the potash industry, aluminium hydroxide, fertilizers, etc.), minerals processing (ore sludge, minerals, coal, etc.) or protection of the environment (flue-gas gypsum, power-station fly-ash, ground remediation, etc.).

Rubber Conveyor Belt Filter The main characteristic of this type of belt filter (Figure 18.22) is an endless rubber belt with drainage channels across it which runs continuously around two reversing rollers to form, together with the filter cloth running synchronised with it, the horizontal filter surface. In the centred longitudinal axis of the belt there are oval openings through which the filtrate running off in the drainage channels reaches the filtrate collecting box below it, which is connected to the vacuum pump. The filtrate collecting box is divided by variable partitions into filtration, washing and drying zones. One of the two rollers acts as the driving roller, whereby the continuous circulation of the rubber belt with the filter cloth is achieved.

The sealing of the stationary vacuum box against the rubber belt is accomplished by means of low-friction slip elements. The rubber belt slides either over a sliding deck, which is fed with air or water to reduce the friction or over a system of low-friction, low-maintenance rollers.

Beyond the drying zone the filter cloth is removed by a system of rollers and on its route back, it is washed on both sides through jet pipes. This guarantees a consistently-high filter capacity. Discharge of the filter cake is carried out on the discharge roller by a sharp reversal of the filter cloth. In this way, it is also possible to remove very thin cakes – with the assistance of a discharge-wire additionally – quite effectively.

On a belt filter the filter cake can be washed in one step or several steps, according to the specification of the end product.

Figure 18.23 shows a typical arrangement for a two-stage counter-current washing washer.

Modular Type Filter With the so-called modular filter (Figure 18.24) the endless rubber carrying belt with its trough shape is replaced by trays of stainless steel

18.3 Methods of Continuous Vacuum Filtration, Types of Design and Examples of Application | 353

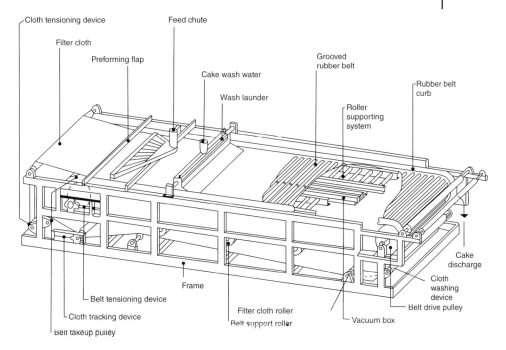

Figure 18.22 The horizontal rubber belt filter.

and either used as standard section or as individual trays to create the horizontal filter surface. We distinguish here between modular filters with stationary filter trays, with indexed tray return and with continuously circulating trays with filter cloth. More on this in [17] and [18].

18.3.3.2 The Pan Filter

Pan filters are continuous vacuum filters in which a horizontal circular surface which rotates around a vertical central axis forms the filtration area. The slurry is fed onto from above, the suction takes place in the direction of the force of gravity. Filtration on a pan filter in relation to the filter surface available offers a space-saving and economical solution for the dewatering of medium- and coarse-disperse systems (ores, crystalline salts, coarser grades of sand, etc.) in widely differing sections of industry.

Pan Filter with Rigid Rim and Screw Discharge, Dorr-Oliver Design Here the active dewatering surface is divided into a number of filter sections by means of swallowtail-shaped division strips. The unit made up of these segments, the filter pan, is supported on a roller-bearing connection mounted in its turn on a supporting framework.

The filter medium is in most cases a filter cloth laid on polypropylene sieve plates or wedged into the dividing grooves; there are however also special versions in use which have stainless-steel sieve type coverage. The filtrate reaches the control

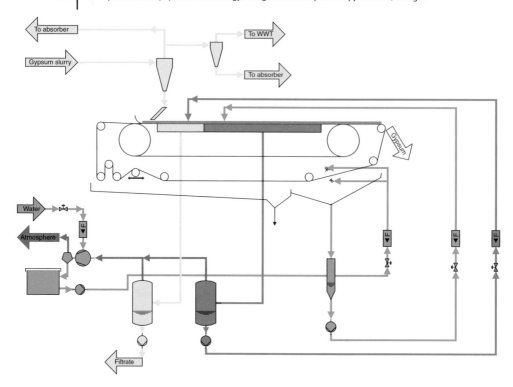

Figure 18.23 Two-stage counter current washing on a vacuum belt filter in a fluid gas desulfurization plant (dewatering of gypsum).

valve through collector pipes. The control valve is arranged below the filter bowl and is forced against the bowl by springs. The entire control valve can be lowered, which enables the control and front plate (wear plate) to be replaced quickly. The discharge screw is mounted floating on a separate stand. Figure 18.25 shows the individual components of a pan filter with rigid rim and discharge screw.

The slurry feed is equally distributed along the whole width of the filter. When the filter pan starts to revolve, the individual sequences of the filter cycle operate in accordance with the angular specifications in the control valve. The mother liquor and as appropriate the wash liquor are drawn through the filter medium by the vacuum applied and the filter cake starts to form or is dewatered.

Once the cycle is completed, the filter cake is discharged via a screw. Where the solids are abrasive, the screw will be fitted with replaceable tip elements. The solids remaining on the filter will be swirled up by a blowback of compressed air and mixed with the newly fed slurry. The rotation speed of the filter pan is generally variable.

Pan filters are predestined for use in counter current washing with precise separation of mother filtrate and wash filtrate. They can also be completely flooded with washing liquor, which has the effect of a displacement wash. For filtration

18.3 Methods of Continuous Vacuum Filtration, Types of Design and Examples of Application

Stationary filter trays

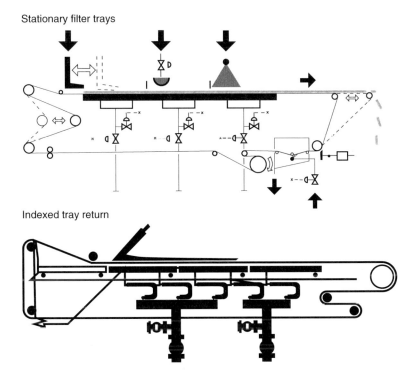

Indexed tray return

Continuously-circulating trays and filter cloth

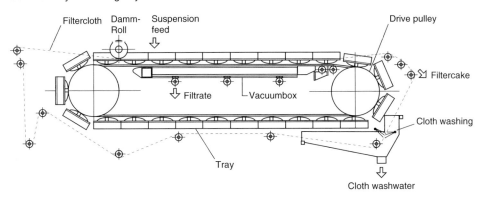

Figure 18.24 Modular filter.

tasks with toxic products, pan filters – due to their circular shape – are easy to enclose to be vapour tight using a hood with vapour seal.

Figure 18.26 shows three pan filters made of mild steel with a filtration area of 62 m² (pan diameter 9.2 m each). The pan filters shown are equipped with two cake wash boxes for counter current cake washing. This pan filter is executed with inclined pan bottom for discharging the filtrate (the filter deck representing the

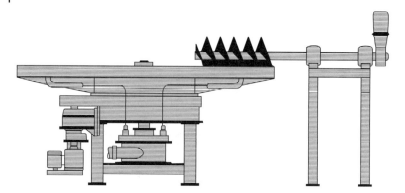

Figure 18.25 Pan filter with rigid rim with screw discharge, Dorr-Oliver Eimco design.

Figure 18.26 Vacuum pan filter, 62 m² active filtration area, manufactured in mild steel.

filtration surface is flat) and is used, for example for the filtration of aluminium hydroxide. Figure 18.27 illustrates a 30 m² pan filter in the work shop with flat pan bottom and externally arranged filtrate pipes manufactured in Hastelloy C-276.

Pan Filter with Tilting Cells, by Bird-Prayon The pan filter with tilting cells (Figure 18.28) has moveable filter cells. These are rotated 180° about an axis radial to the filter on completion of the filtration cycle. Then a hard jolt on the cell together with interruption of the vacuum supply ensures a complete discharge of the filter cake. The filter cloth thus becoming accessible, it can be thoroughly cleaned, which is not a the case on other pan filters. More on this in [18] and [19].

18.3 Methods of Continuous Vacuum Filtration, Types of Design and Examples of Application | 357

Figure 18.27 Vacuum pan filter, 30 m² active filtration area, manufactured in Hastelloy C-276.

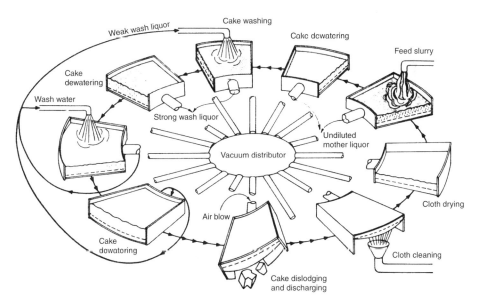

Figure 18.28 Pan filter with tilting cells, by Bird-Prayon.

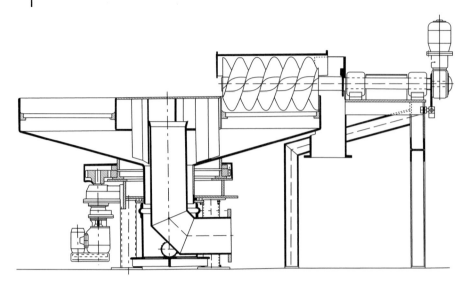

Figure 18.29 Sand pan filter with syphon.

Single-Cell Pan Filter with Syphon, by Dorr-Oliver This type of pan filter (Figure 18.29) was developed for the dewatering of fine-grain sand. This is a single-cell pan filter, which has no need of a control valve. The required difference in pressure – mostly only about 200 mbar – is not generated by vacuum pumps, but by single- or two-stage blowers. Further details under [18] and [20].

Pan filter with Loose Edge and Screw Discharge, by Ucego This joint development by the Messrs. Pechiney Saint-Gobain and Union Chimique Belge has as its characteristic hallmark a rubber belt, which runs around the outside of the filter plate and serves as a seal against the escape of slurry. The rubber belt is diverted in the vicinity of the discharge of the filter cake by means of rollers, this providing space for the filter cake to be discharged away from the plate. It was possible in this way to avoid the patented lifting screw discharge of the Dorr-Oliver design. Further details under [18].

18.3.4
Vacuum Filtration on a Vertical Flat Surface, the Disc Filter

In this type of filter (Figure 18.30), the filtration is carried out at right angle to gravity using several parallel discs mounted on a hollow horizontal shaft. Two important advantages enjoyed by this design, among others, are the accommodation of a large filter area in an extremely restricted foot print and the possibility of increasing the filter area by one or more discs if this should prove necessary for process engineering reasons. Disc filters are particularly suited to dewatering

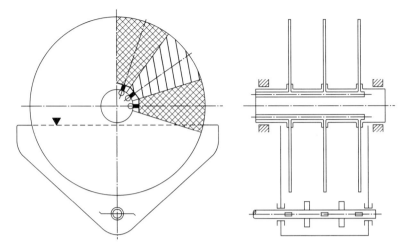

Figure 18.30 Disc filter.

slurries in which the grain size differences in the solids are not too extreme – in other words, the smallest particles must still enable a dischargeable filter-cake thickness to form. In the case of slurries including coarser solids, a correspondingly fine-grained proportion must be available.

The filter discs consist of a number of sectors. Their number is dependent among other things on the diameter of the discs. The individual sectors have a grooved and/or perforated surface, which is covered both sides with the filter cloth. The discs rotate in the disc-filter trough and during one revolution, they go through one complete filtration cycle. Via the filtrate pipe system accommodated

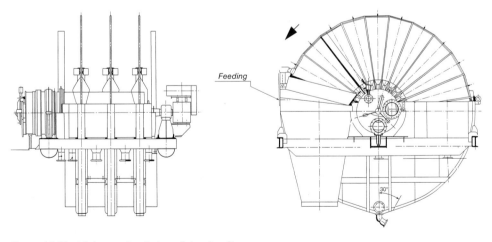

Figure 18.31 High-capacity design of the disc filter.

in the hollow shaft. Filtrate from the individual sectors reaches the control head, which apart from two suction connections also has a blowback connection. A compressed air blowback is led from here into the filter-cake discharge zone. The cake falls over the discharge scrapers fitted on either side of each disc into the discharge chute.

To prevent certain slurries from segregating, the disc filter can be fitted with a paddle-type mixer or an air agitator [21]. Apart from this, the filter trough can be divided into a number of separate chambers, so that the filter can be fed with different sludges at the same time. If due to the nature of the filter cake a discharge by scraper, is not feasible disc filters with roller discharge can be used.

Disc filters have proved themselves in the filtration of mining slurry, fire coal, coal-washing water, raw cement slurry, blast-furnace slurry, aluminium hydroxide, as well as flotation concentrates. Figure 18.31 shows the high-capacity design used for aluminium hydroxide filtration with single troughs (one trough per disc) without agitator. The slurry is kept in suspension by rotation of the discs. Figures 18.32 and 18.33 illustrate a production filter of this type with three discs, for dewatering aluminium hydroxide.

Figure 18.32 High-capacity or large diameter disc filter in operation, filtration area 120 m^2.

Figure 18.33 Filter-cake discharge by blowback.

References

1. Redeker, D. (1987) Strategien für Trennapparateauswahl und −auslegung. Preprints GVC-Tagung Mechanische Flüssigkeitsabtrennung, Köln, Germany, 30 November/1 December, 1987.
2. Lahdenperä, E., Korhonen, E., and Nyström, L. (1989) An expert system for selection of solid/liquid separation equipment. *Chemical Engineering*, **13**, 467–474.
3. Lesage, G. and Chacar, R. (1990) in *An Expert System for the Design, Operation and Diagnosis of Solid/Liquid Separators, Computer Applications in Chemical Engineering* (Hrgs. H.T. Bussemaker and P.D. Iedema), Elsevier.
4. Carman, P.C. (1938) Fundamental principles of industrial filtration. *Transactions of the Institution of Chemical Engineers (London)*, **16**, 168–188.
5. Chalmers, J.M., Elledge, L.R., and Porter, H.F. (1955) Filters. *Chemical Engineering*, **62** (6), 191.
6. Uhlemann, J. (1997) Rechnergestütze Problemlösungsstrategien in der mechanischen Flüssigkeitsabtrennung. *Chemie Ingenieur Technik*, **69**, 765–775.
7. Dr.-Ing. Nicolaou, I. (2003) Kuchen bildende Filtration von Suspensionen und Filterberechnung. *Chemie Ingenieur Technik*, **75**, 1–15.
8. Dorr-Oliver GmbH Wiesbaden. Systemlösungen für die Filtration, Firmenprospekt.
9. Dorr-Oliver GmbH Wiesbaden. Filter F70, Firmenprospekt.
10. Schwalbach, W. (1980) *Zur Bedeutung des Filtratrohrsystems bei Trommelfiltern, Sonderdruck 9*, Hüthing Verlag, Heidelberg, S. 569–571.
11. Hackl, A. and Höflinger, W. (1984) Zur Modellierung der Anschwemmfiltration am Vakuumtrommelfilter. *Chemische Technik*, **36** (1), 23–29.
12. Smith, G. (1976) How to use rotary vacuum precoat filters. *Chemical Engineering*, **83** (4), 84–90.
13. Dorr-Oliver GmbH Wiesbaden. Entparrafinierungsfilter, Firmenprospekt.
14. Zogg, M. (1987) *Einführung in die Mechanische Verfahrenstechnik*, B. G. Teubner, Stuttgart, S. 128.
15. Dorr-Oliver GmbH Wiesbaden. Kompaktfilter, Firmenprospekt.
16. Schwalbach, W. (1991) in *Maschinen + Apparate zur Fest/Flüssig-Trennung*, 1

Ausgabe (Hrsg. W.F. Hess), Vulkan-Verlag Essen, S. 332–337.
17. Zogg, M. (1987) *Einführung in die Mechanische Verfahrenstechnik*, B. G. Teubner, Stuttgart, S. 122.
18. Schwalbach, W. (1991) in *Maschinen + Apparate zur Fest/Flüssig-Trennung*, 1 Ausgabe (Hrsg. W.F. Hess), Vulkan-Verlag Essen, S. 210.
19. Ullmanns Encyklopädie der technischen Chemie (1972) *Verfahrenstechnik I (Grundoperationen)*, Bd. **2**, Verlag Chemie, Weinheim.
20. Dorr-Oliver GmbH Wiesbaden. Sandplanfilter, Firmenprospekt.
21. Artelt, H. (1990) in *Handbuch der industriellen Fest/Flüssig-Filtration* (Hrsg. H. Gasper), Hüthing Verlag, Heidelberg, S. 162–164.

Index

a

activity coefficient   21
air
– -flow rate   317ff
– –steam mixture   19f
– –vapour mixture   18f
Amonton's law   6
Avogadro's law   6

b

back-mixing   222
backflow   113, 124
backing pump   107, 114, 116, 287
barometric
– installation   30ff.
– temperature measurement (BTM)   265f.
belt filter   352
boiling
– point   138, 225f., 282
Boltzmann constant   3f.
Boyle–Mariotte law   6, 163

c

calcareous deposits   75ff.
cavitation, see damage
centrifuges   203, 208
Charle's law   6
chemical process engineering   128f., 295, 313
clean-in-place (CIP)   275
CLP Ordinance   138
coating   222, 225, 230
cold trap outlet   288
collecting tank   30f.
combustible substances   130, 132, 134, 140f., 145
compression
– dry   104, 118
– energy   44
– heat   108
– inner   119f., 122f.
– maximum   115
– over-   119
– pre-   69ff.
– pressure   45
– ratio   90, 105
– re-   227
compressor
– claw   111, 119
– displacement   36
– double-sided pressurized liquid ring   37f.
– gas jet   86ff.
– liquid ring   36ff.
– rotary piston   69, 72
– single-sided pressurized liquid ring   37, 39
– steam jet   82, 85ff.
condensate
– circulation   33
– film   27
– hot   222
condensation   15f., 28f., 283, 295
– heat   15
– inert gases   16f.
– plant   33
– process   18, 26
– specific enthalpy   15f.
– surface   18, 23, 91
– temperature   21 2f., 28f.
– vapour   27
condenser
– air   24
– block   24, 92
– co-current   22f.
– counterflow   22f.
– direct-contact   21f., 33
– downstream   69, 107

Vacuum Technology in the Chemical Industry, First Edition. Edited by Wolfgang Jorisch.
© 2015 Wiley-VCH Verlag GmbH & Co. KGaA. Published 2015 by Wiley-VCH Verlag GmbH & Co. KGaA.

condenser (cont.)
- exhaust emission   122
- floating head   24, 26
- ice   266, 271
- installation   30ff.
- mixing   23, 91
- pre-condenser   61
- pressure   87
- spray   22
- surface   19, 23f., 92
- tubular-type   23f.
- U-tube   24
- vacuum control   30
contamination   222, 275
continuous filtration   345
- vacuum drum filter   345
cooling
- compressor   273
- down rate   260
- open circuit   69
- pre-admission   117
- rate   272
- water   87f., 92, 95
- water temperature   282
corrosion
- liquid ring vacuum pump   77f.
- steam jet vacuum pumps   82
- vacuum evaporation   215
- vapour   104
critical
- expansion ratio   9
- pressure ratio   9
crystal
- growth   193f.
- growth rates   192, 195
- habit   189
crystal size distribution (CSD)   189, 195
crystallisation process   189f., 203, 206ff.
- evaporation   199, 207
- salt   199
- sodium chloride   207f.
- vacuum cooling   193, 197, 206f.
crystalliser   192ff.
- double-propeller (DP)   201
- draft-tube-baffled (DTB)   192, 196f., 201f.
- fluidised-bed type   201, 203f.
- forced-circulation (FC)   192, 196ff.
- horizontal   197
- Oslo-type   192, 196, 201
- production caoacity   193
- stirred-tank   197
- turbulence (MESSO)   201
- vertical, agitated-tank   197

d

Dalton's law   3, 42, 161
damage
- effects   145f.
- liquid ring vacuum pump   73ff.
- steam jet vacuum pumps   82
dead spaces   54f.
defrosting   286
degassing   99
diffuser   82, 84
distillation, *see* molecular distillation processes
disc filter   359
discharge   339, 341
- chain or string   342
- cloth   341
- precoat   342
- roller   340
- scraper   339
Dorr–Oliver
- pan filter   358
downstream pump   111
drum filter   337
dryer designs   247f.
drying   99, 235ff.
- batch system   244ff.
- capacity   241f., 246
- contact   236, 238f., 242
- continuous vacuum   246ff.
- convection   236f.
- curve   246
- discontinuous vacuum   244ff.
- parameters   245
- process sequences   236
- radiation   236f.
- sun-   237
- under vacuum   241f.
- vacuum lock systems   247

e

ejector   84, 86, 89, 93
- downstream   90
- gas   69ff.
- multi-stage   92, 321
energy conservation laws   41
environmental protection   251
equation of state
- ideal gas   6f., 14
equilibrium, *see* phase equilibrium
equivalent pipe length   10f.
European directive 94/9/EC   150ff.
European directives for explosion protection 99/92/EC   131, 144, 146ff.
evacuation process   12f.

evaporation
- capacity   240, 242, 244
- –crystallisation process   199, 207
- ideal vacuum   212
- pressure   212f.
- process   16, 190
- rates   251
- temperature curve   241f.
- thermodynamics   211ff.
- thin film   99
evaporation plant   213, 218f., 222
- cost efficiency   223
evaporator
- agitator   223f.
- batch   290
- climbing-film   225, 227f., 231
- falling-film   223, 226, 228f., 290
- fluidised-bed   230
- forced-circulation   228ff.
- HADWACO system   232
- multi-effect   227
- natural circulation   223, 225, 228f., 290
- plate   231ff.
- rising tube   223, 226
- short path   285f., 290
- wiped film   284, 286, 290
explosion
- -proof equipment   129, 176
- protection   127, 129ff.
- range   133, 135
explosion group (Ex-Group)   133, 145
explosion limit
- lower (LEL)   133ff.
- upper (UEL)   133ff.
explosion point
- lower (LEP)   136f.
- upper (UEP)   136

f

filter cake   339
- discharge   339
filter sizing software (FILOS)   332
filter type selection   335
flame
- arresters (FAs)   141f.
- -proof gap width   141
flammable liquid   136ff.
flash point   136ff.
floating head   24, 26
flow
- co-current   22
- continuous   7
- gas   9, 317
- laminar   8, 11

- mass   8
- molecular   8, 11
- motive   88
- plugged   25
- pressure   256
- supersonic   85
- turbulent   7f., 231
- vapour   14, 24, 240, 245f.
- viscous gas   7
foaming   222
fouling   29f., 34, 82, 282, 298
fractionation split   299
freeze-drying process   32, 259ff.
- advantages   259
- chamber   265ff.
- chamber-condenser valve   270f.
- cleaning   275f.
- final treatment   264
- primary drying (sublimation)   261
- process control   265f.
- refrigerating system   271
- secondary drying   264
- shelf-assembly   267ff.
- stoppering process   264
- vacuum system   273f.
freezing point   259ff.
frequency-controlled motor   112, 126
friction
- heat   62
- losses   41
fugacity coefficients   298

g

gas
- ballast   106f.
- jet   69ff.
- load capacity   303f.
- separation   52, 296
- throughput   8f., 12
- –liquid mixture   57
- –vapour mixture   19, 22, 29
- velocity   10
Globally harmonised system (GHS)   138

h

hazardous areas   149f.
heat
- conductivity   262
- energy transfer under vacuum   256
- exchanger   61, 206, 231
- flow   24, 27, 238, 242
- input   44
- resistance   239
- sensitive products   283, 290

heat transfer   212, 222, 224, 263
– area   24, 238
– coefficient   23f., 29, 231, 238
– contact drying   237
– controlled   240
– liquid   237
– radiation drying   237
– surfaces   222, 230
heat transition   225f.
heating
– re-   261
– steam   212, 214
HETP (hight of packing equivalent to one theoretical plate)   303ff.
highest acceptable risk (R_v)   145
hub
– conical   39
– control   54f.
– cylindrical   39
hydrodynamics   305
hydrostatic manometers   163

i
ignition
– chamber   142
– energy limiting curve   139
– pressure   256
– probability   146
– source   136f., 139
– temperature   138f., 256
impeller   39f., 108, 118
– cavitation   73
– hub   41
– slots   101
– suspended   101
incrustations   34f., 206, 228, 233
inert gas   16f., 27
– fraction   29
– saturated   19
inlet
– opening   40
– pressure   60
– velocity   26
inter casing   39
– cavitation damage   75
isobar   6
isochore   6, 103

j
jet
– gas   81
– liquid   81
– motive   84

k
kinetic energy
– gas mixture   70
– vacuum fluid bed   255
kinetic gas theory   3, 5
Knudsen number   7f.

l
leak detection
– high-vacuum range   177f.
– limits   179
– medium-vacuum range   177f.
– methods   173, 175f.
– rough vacuum range   177f., 179, 183
– signal response time   181f.
leak detector   176ff.
– helium   176ff.
– specifications   184f.
– WISE technology   185
leakage
– gas flows   14
– rates   173f., 179, 182
lower explosion limit (LEL), *see* explosion
lubrication   40, 100
– circulatory   102f., 104, 110
– gear-oil   125
– self-   111

m
Mach number   94
magnetic coupling   62f.
manometer
– Bourdon   164
– diaphragm   164
– hydrostatic   163
– U-tube   163
mass flow rate   19, 84, 86f.
mass transfer
– efficiency (HETP)   304
– equipment   296, 303
maximum experimental safe gap (MESG)   142ff.
McCabe–Thiele methology   297f.
mean free path   5, 7, 284
microbe growth   221f.
minimum ignition energy (MIE)   132, 134, 138ff.
Modular filter   355
moisture removal   239, 241
molecular distillation processes   12, 99, 281ff.
– applications   290ff.
– column   12f.
– plant   289f.

Index | 367

– pump 282
– short path 283ff.
– system 177, 179
– vacuum distillation 282
Mollier enthalpy–entropy diagram 83
Montz high-capacity packing 306ff.
motive
– media 93, 95
– nozzle 84f., 87, 94
– steam 82, 90
– vapour 89, 94
multi-stage steam jet vacuum 90ff.

n
nozzle
– ejector 70
– Laval 69, 84, 319f.
– mixing 84f.
– needle 88f.
– pressure 82, 222
– propellant 71
– suction 82
– supersonic 84, 320
– vacuum 287
nucleation
– kinetics 191f.
– rate 193, 196f.

o
Ohm's law of vacuum 8, 10
operating liquid 43, 68

p
pan filter 353, 356, 358
– Bird-Prayon 356
– Dorr-Oliver 358
– Ucego 358
particle number density 3, 5
petrochemical industry 295
pharmaceutical process engineering 128, 221, 266ff.
phase diagram
– solvent 261
– water 261f.
phase equilibrium
– steam–water 18
– vapour–liquid 20, 297f.
phase transition 236
port cylinder 55f.
power consumption 40, 45f., 60, 117f.
pressure
– absolute 161f.
– atmospheric 16f., 111, 134f., 161f.
– backing 107, 114f.

– compression 45
– constant 6, 16
– control 63
– counter- 38, 45
– cut-in 116
– differential 161, 256
– discharge 90, 85f.
– drop 226ff.
– flow out 9
– gauge 161f.
– gradient 240
– loss 226, 319
– loss coefficient 41
– maximum 115f.
– maximum explosion 134f.
– motive 85f., 88
– motive steam 90
– operation 90
– outlet 81
– over- 162, 164, 214
– partial 3, 18, 20, 28, 105, 136, 161
– plant 214
– ports 37f., 40, 53
– range 2, 52, 100
– ratio 84
– sensor 165f.
– spraying 255
– standard 7, 9, 16f.
– supercritical 84
– total 3, 20, 105, 163, 183, 261
– upstream 116
– working 99f., 183, 221
pressure measurement
– direct 161f.
– indirect 162
propellant
– gas 71
– nozzle 71
pump-down processes 10, 13
pumping speed 12f., 110, 113, 117, 288
– diagram 12f.

r
reboiler 295
rectification 282, 295ff.
– applications 309
– design 302f.
– ethanol 300f.
– pressure 298
– structured packings 305ff.
– vertical columns 282, 295, 299f.
redistributers 296

Index

reflux 295, 299
residence time 222, 241, 304
Reynolds number 7f., 315
rinsing 222
roller discharge 340
roots blowers 111

s

safety
– absolute 131
– characteristics 132
– regulations 131, 144, 146f., 251
– technical requirements 129ff.
seals
– double-acting mechanical 62
– single-acting mechanical 59, 62
separation
– efficiency 303f.
– noises 57
– solid–liquid 235
– trays 303
separation techniques 331
– mechanical 331
solid deposits 54
solution 20f.
specific gas constant 45
steam jet vacuum pump 71f., 81ff.
– design 81ff.
– function 81ff.
– liquid ring 93
– types 83
sterilization
– cold 277
– steam sterilization cycle 276f.
– VHP (vaporized hydrogen peroxide) 277f.
still 282ff.
strain gauges 165f.
stripping section 296f.
sublimation 99, 259ff.
– zone 263
suction
– capacity 42f., 55, 60, 85f.
– flow 64
– hydraulic 42, 44
– power 64
– pressure 74, 81, 86, 88
– speed diagram 113
– volume 41
supersaturation 190ff.
– control 193
– cycle 194
surge tanks 50

t

temperature
– absolute 3
– class (T-class) 133
– constant 16, 40
– coolant 28
– ice 262
– runoff 28
– standard 7
thermal
– degradation 298f.
– efficiency 243
– product treatment 242
thermodynamic 41
– equilibrium 295
throttle control valve 88, 288
throttling 224, 233
– orifice 230
tracer gas 176f.
tube
– bundle 24
– fixed 25
– heat-exchanger 33
– horizontal 26f.
– vertical 26

u

upper explosion limit (UEL), see explosion

v

vacuum
– crystallization 99, 189ff.
– distillation, see molecular distillation processes
– dryer, see dryer designs
– see drying
vacuum control 63f.
– condensers 30
– liquid ring vacuum pump 63
vacuum conveying 311ff.
– applications 312f.
– continuous 325f.
– dense phase 315ff.
– design 317f.
– dilute phase 315ff.
– filter systems 324ff..
– loading 325ff.
– MULTIJECTOR 325, 327, 330
– parameters 329
– plug flow 315, 317
– pneumatic 314
– positive pressure 312, 314
– receiver 322
– vessels 322

vacuum drum filter   346
vacuum filtration   331, 337, 345, 351, 352, 358
– belt filter   352
– continuous   337
– curved concave surface   351
– curved convex surface   337
– disc filter   358
– drum filter   337
– flat horizontal surface   352
– internal filter   351
– pan filter   353
– vertical flat surface   358
vacuum filtration plants   335
vacuum fluid bed   251ff.
– applications   256f.
– closed-loop system   253
– components   253f.
– control system   251
– investment costs   253
– once-through units   251f.
– operation   255f.
– process   251ff.
vacuum gauges   161ff.
– Bourdon tube   164
– calibration curves   170, 172
– compression   163
– diaphragm   164ff.
– directly   162
– electromechanical   164
– indirectly   162
– mechanical   164
– mercury   161
– offset   170
– thermal conductivity   167ff.
vacuum pumps
– adsorption   12
– air-cooled screw   127
– cascade connection   64ff.
– claw-type   119ff.
– compression backing   110f.
– diaphragm pumps   100f., 110
– diffusion   12
– dry compression   104, 111, 120
– fresh-oil lubricated   108f.
– high-vacuum   12
– horizontal screw   127
– jacket-cooled three-stage roots   118
– kinetic   12, 100
– kinetic gas   12
– liquid ring   35f., 100
– lobe-type   103
– mechanical   97ff.
– multicell   100, 108

– multi-stage steam jet   90ff.
– multi-vane   111
– oil-free compression   110
– oil-sealed rotary vane   104, 111, 164
– piston   100f.
– positive displacement   12, 99
– reciprocating piston   100
– roots   100, 102f., 110ff.
– rotary piston   72, 100, 102f.
– rotary plunger   102, 104
– rotary vane   12f., 100ff.
– screw-type   124ff.
– set-up   67f.
– single-stage closed coupled   57f.
– single-stage liquid ring (LRVPs)   37, 41ff.
– speed adjustment   65f.
– speed-controlled   66
– steam jet   81ff.
– turbomolecular   12
– two-stage   57f., 61, 110
– valve control   63
  vapour recovery   63
vacuum range
– medium   12
– rough   12
valve
– bypass   112f., 116
– control   63, 88
– non-return   118
– technology   54
vapour
– concentration   136
– density   284
– distributers   296
– recovery   63
– tolerance   104ff.
– utilization   217ff.
vapour pressure   57, 136, 214
– diagram   15f., 20
velocity
– gas   4f., 257, 303
– sonic   84
venting processes   10, 31
vibration damping   50
vibration simulation
– flow separations   50
– liquid ring vacuum pumps   48ff.
– pulsation   48f.
viscosity   223
– kinematic   315
– liquids   230
– solids   290
– water   216
volatilities   298

volume
- damage 43
- flow rate 91, 273
- throughput leakage 173f.

W

water
- for injections (WFIs) 275
- impact, *see* damage
- quality 76f.
- steam compression 220
- vapour pressure 18, 262
- waste 82

working
  chamber 41, 43
- openings 39f.

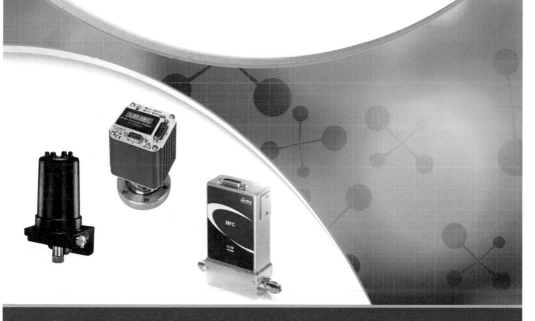

Vakuum
in Forschung und Praxis

The expert journal for vacuum and plasma technology and thin films

Vakuum in Forschung & Praxis is the expert journal for vacuum and plasma technology and thin films. This editorial focus is unique in German-speaking countries and sees itself as the link between scientists, practitioners and users.

Vakuum in Forschung & Praxis (VIP) publishes overviews, essays, refereed articles from research and user reports. The regular rubrics "Vacuums for Users", surface analytics, product information, and "People" as well as announcements and reviews of trade fairs and conferences round off the content. VIP appears six times a year.

WWW.VIP-JOURNAL.DE

For subscription details please contact Wiley Customer Service:
- cs-journals@wiley.com (America, Europe, Middle East and Africa, Asia Pacific)
- cs-germany@wiley.com (Germany, Austria, Switzerland)

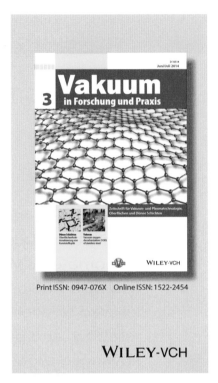

Print ISSN: 0947-076X Online ISSN: 1522-2454

WILEY-VCH

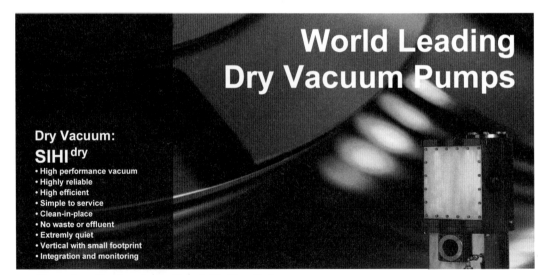

World Leading Dry Vacuum Pumps

Dry Vacuum:
SIHIdry
- High performance vacuum
- Highly reliable
- High efficient
- Simple to service
- Clean-in-place
- No waste or effluent
- Extremly quiet
- Vertical with small footprint
- Integration and monitoring

Pumping Technology For A Better Future

SIHIdry is a vertically oriented and self draining vacuum pump with no mechanical shaft seals. It is an ideal choice for chemically related processes where there is a high possibility of liquids or solids carry-over.solution.

Sterling SIHI GmbH - www.sterlingSIHI.com